FLUIDIZED BED COMBUSTION
OF COAL AND WASTE MATERIALS

FLUIDIZED BED
COMBUSTION OF COAL
AND WASTE MATERIALS

Lee Yaverbaum

NOYES DATA CORPORATION

Park Ridge, New Jersey, U.S.A.

1977

Published in the United States of America by
Noyes Data Corporation
Noyes Building, Park Ridge, New Jersey 07656

FOREWORD

Fluidized bed combustion is an ingenious, low cost method for obtaining energy from high-sulfur coal in an environmentally acceptable manner. Feasible processes are described here in detail.

The technology can also be used to substitute many low grade fuels directly for coal as outlined in the introduction. Energy contributions from low temperature fluidized bed combustion will probably be substantial in the future.

The practicability thereof is amply demonstrated in this book which is based on government-funded studies and important U.S. patents.

Advanced composition and production methods developed by Noyes Data are employed to bring our new durably bound books to you in a minimum of time. Special techniques are used to close the gap between "manuscript" and "completed book." Technological progress is so rapid that time-honored, conventional typesetting, binding and shipping methods are no longer suitable. We have bypassed the delays in the conventional book publishing cycle and provide the user with an effective and convenient means of reviewing up-to-date information in depth.

The table of contents is organized in such a way as to serve as a subject index and provides easy access to the information contained in this book. The bibliography at the end of the volume lists the highly important government reports and a source of their purchase.

CONTENTS AND SUBJECT INDEX

PART II.   WASTE HANDLING IN FLUIDIZED BEDS

RESIDUAL OIL GASIFICATION/DESULFURIZATION

SEWAGE, INDUSTRIAL AND OTHER WASTE TREATMENT
PROCESSES

INTRODUCTION

Coal represents approximately 85% of the U.S. fossil fuel resources and will certainly play an important role in achieving greater energy independence. Of all our fuel reserves, it merits the technical development of solutions to the problem of increasing the efficiency of thermodynamic cycles and to pollution problems associated with its use. There are three pollution problems of basic concern—sulfur emissions, nitrogen oxide emission, and excessive heat rejection to rivers.

Fluidized Bed Combustion (FBC) has a great number of attractions for getting energy from coal—it is a more compact, cheaper method of burning high-sulfur coal and other fuels in an environmentally acceptable manner, thereby increasing the world's inventory of usable fossil fuel resources. In the basic concept, coal, which is crushed but not ground, is fed into a hot bed of dolomite or limestone that is fluidized by hot air. Water, pumped through coils immersed in the bed, is converted into steam for electricity, process heat or other application. Temperature in the reactor is maintained at 1500° to 1700°F (800° to 950°C), instead of the 2500° to 3000°F of regular boilers.

The sulfur in the coal comes off as sulfur dioxide, which is captured by the sorbent (dolomite or limestone) and converted into calcium sulfate. The lower temperatures decrease steam tube corrosion, allow the use of lower grade coals (since these temperatures are lower than ash slagging temperatures), and also decrease NO_x emissions; they also prevent ash agglomeration: minerals exit overhead as fly ash that is collected in cyclones.

Operation at elevated pressures, in the range of 600 to 1,000 kPa, offers further advantages—the hot flue gas from a pressurized system can be expanded through a gas turbine, thereby increasing the power generating efficiency even further. Pressurized fluidized bed combustion for electric power generation therefore provides a direct combustion process for coal, with the potential for improved thermal conversion efficiency, reduced costs, and acceptable environmental impact.

This technology can be used to substitute coal and other low-grade fuels for oil and gas and to provide an improved option for conventional coal-fired systems,

1

that is, it has the capability of burning many types and grades of coal, including tailings, as well as municipal sludge and refuse, industrial and agricultural cellulosic waste materials, oil shale and petroleum fractions. With successful development of the technology, contributions from fluidized bed combustion technology to the energy system have been projected to be up to 40,000 MW of installed capacity by the year 2000.

Part I
Coal-Fired Fluidized Beds

DEVELOPMENT OF COAL-FIRED FLUIDIZED BED BOILERS

The material in this chapter was excerpted from PB 234 343 and *Technology Ireland.* For a complete bibliography, see p 263.

BACKGROUND

History

Although the fluidized bed boiler appears to be new and novel, it actually evolves logically out of the prior art of the steam power engineer. The first major class of boiler system for solid fuels, the fixed bed, dates from antiquity. To meet the needs of the industrial revolution the "mechanical fireman," the stoker, was devised. The first patent in this area was Watt's in 1785. Almost simultaneously it was recognized that if coal were finely divided, i.e., pulverized, it might be burned in a suspension similar to a gas. But suitable techniques for pulverizing were not yet available. Not until the 1890s did pulverizing methods and incentives appear in the cement industry and in 1921, the first significant application of pulverized coal firing to a utility boiler was made. From that beginning, this form of combustion came to dominate the utility market, growing by a factor of 100 within ten years.

In that same year, 1921, Fritz Winkler had devised a system for the manufacture of fuel gas. In his United States patent, Winkler describes his system thus: ". . . small-sized coal . . . (charged) . . . onto a suitable grate . . . and blowing a gasifying medium, such as air, . . . with sufficient pressure to bring about a strong agitation in the form . . . of a boiling motion of the material. On the other hand the blast of gas should not be so strong as to carry away the solid fuel and blow it out of the generator. A free passage of the gases is secured . . . the material is thoroughly mixed . . . its combustion is not concentrated locally but well distributed . . . a uniform temperature being thereby established . . . . Each particle of fuel is brought into intimate contact with the gas blown in from below, but not so much gas that an undue removal of heat and dilution . . . would be caused. It is characteristic of my invention that the entire bed of fuel is kept in continuous agitation by means of a gas current blown through the charge . . . and that this state of continuous agitation is maintained merely by the gas current and without any mechanical means."

This description makes clear that Winkler had invented the fluidized bed combustor. Had he added suitably arranged heat transfer surfaces, he would simultaneously have invented a fluidized bed boiler. One additional oversight—Winkler chose to name the device but not the process. His system, known as the "Winkler Gas Generator," supplied much of Germany's raw materials for synthetic gasoline in World War II.

The term "fluidization" came into wide use in the petroleum industry in the 1940s. In 1944, a patent application was filed by a petroleum researcher for a coal fired, fluidized bed boiler. This early concept was flawed by the inventor's fundamental misconceptions about what is feasible in a boiler design, and the development was not pursued.

The current state of the art of fluidized bed boilers owes little or nothing to that previously developed for catalytic cracking.

The Need for a New Form of Combustion

If Winkler's work on fluidized bed combustion had come 30 years earlier, the fate of pulverized fuel firing would have been different—stokers had seen their day in central stations and a new system was required. A close analysis of how technology evolves shows that not much would have changed. The first applications of pulverized fuel firing to boilers were made by modifying stoker fired units. The first units built solely for pulverized fuel firing retained essentially the same configuration as boilers of that day. This was, essentially, a refractory chamber or furnace in which the fuel was burned and a boiler (set almost apart from the furnace) in which the steam was generated.

The fluidized bed combustor, not boiler, could have been applied to such a configuration. The fluidized bed boiler, however, involves fundamental changes in the configuration of the entire steam generating system.

There are several factors which today lead to the acceptance of the required change. First is the size and cost of today's coal fired boilers. Second is the need to reduce air pollution from coal fired boilers. And third is the need to be able to burn char and other solid fuels. There are two classes of boilers to which this concept may be successfully applied to reduce size and cost—the large industrial boiler and the utility boiler.

Existing coal fired industrial boilers have been hampered by the inability to achieve a packageable design. Oil and gas fired boilers capable of producing 200,000 pounds per hour of steam can be assembled in a factory and brought to the user's site, whereas coal fired boilers must presently be delivered in pieces and assembled at the site.

In the utility field, boiler sizes have increased to match the ever increasing generating unit size. A boiler built to deliver steam for a 1,000 megawatt (MW) power station can no longer be built along the accepted lines of a single waterwall enclosure.

Boiler designers have found it necessary to build such a unit as two or even four chambers. In one instance, each furnace chamber is 34 feet by 64 feet in plan and 138 feet high.

Pulverized fuel firing reached the limit where it offered simplicity when units of 800 MW size were reached. It is difficult to properly project the fuel into a single chamber of the proportions required for a 1,000 MW furnace. This is the same factor, fuel injection, which limits the spreader stoker to less than 100 MW.

The factors of excessive cost and size might not in themselves be sufficient reason to depart from the tried and true methods of burning coal. However, the requirements for reduced emissions of sulfur dioxide (SO_2) are difficult to meet simply and effectively with the conventional forms of combustion.

While the fixed bed of the stoker limits particulate emission potential, it does not permit intimate mixing with a sulfur absorbing additive which might be effective. The pulverized fuel flame does offer the intimate mixing, but the temperature within the flame itself does not permit the absorption of sulfur dioxide by any known additive. The elimination of sulfur emissions must be accomplished by methods which, by necessity, are added to the system in a manner which increases both size and capital cost.

The characteristics of fluidized bed combustion, in general, and of the fluidized bed boiler, in particular, are such that it is possible to limit the emissions of SO_2 by essentially simple techniques. In two boiler markets—industrial and utility—and for two reasons—economy and pollution control—the time for this form of combustion and steam generation has arrived.

The need for such a development became evident in the United States in the 1950s. However, it was in the 1960s that both in the United States and in England a commitment of adequate resources to highly innovative organizations was made to pursue the development.

UNDERSTANDING THE FLUIDIZED BED BOILER

To appreciate the advantages of the fluidized bed boiler, an understanding of how this system differs from conventional boilers is necessary. Although for most the two terms "fluidized bed" and "boiler" might be well understood, a review of these terms is helpful.

Description of a Boiler

A boiler is a system in which a fuel is burned and steam produced. There is no difference in function between a fluidized bed boiler and a conventional boiler or even a nuclear boiler. In all boilers, excluding nuclear, a fuel is burned with air and the heat released by this chemical reaction is used to heat, vaporize and superheat a working fluid which almost universally is water. The resulting steam may then be used to generate electricity via a turbine generator, or to provide heat for some industrial process.

The region in which the fuel is burned is termed the furnace, and in a coal fired boiler the furnace serves three functions: (1) reaction chamber, (2) cooling chamber and (3) ash disposal. Because the furnace serves the three functions simultaneously, a number of design compromises are required in conventional practice. The techniques to promote one function conflict with those used to promote another. Using pulverized fuel firing as an example,

(a) Efficient burning of the fuel is promoted by high temperatures—
cooling the gases is in conflict with this function.

(b) Cooling the gases is promoted by cooling surface being evenly
spaced through the volume of the furnace; however, burning
in a suspension flame is impractical with this most efficient
tube arrangement.

(c) Preparing ash for collection is promoted by firing fuel in relatively
large pieces; efficient combustion is promoted by fine grinding.

(d) Collecting ash before it leaves the furnace would be promoted
by causing the molten ash particles to adhere to the tubes and
flow as a slag to a collection point, but tubes covered with a
molten slag are not effective for cooling gas.

In the fluidized bed boiler, certain design compromises are also necessary but they are of a different nature, and in the areas of complete combustion, gas temperature preparation and ash collection there are no fundamental conflicts.

Description of a Fluidized Bed

As previously stated, the fluidized bed boiler is a derivative of work by Fritz Winkler carried out before 1921, while the term "fluidized bed" was coined in the 1940s in the petroleum industry. The selection of this name indicates the aspects of the system which appeared most significant to the petroleum industry. The name fluidized bed would not have been chosen by Winkler nor by those who might have developed a boiler using Winkler's principles.

The derivation of the term fluidized bed arose from work on catalytic cracking. At one time cracking was practiced in a fixed bed, which is analogous to the stoker, and then in a dilute phase reactor, which is analogous to the pulverized fuel system. Neither of these systems met the needs of the petroleum industry, and by a very costly process of trial and error, the fluidized bed reactor was developed.

It was noted that if a bed of finely divided catalyst was supported on a properly designed grid, and if the petroleum vapors passed up through the bed at sufficient velocity, the particles were buoyed upward and would begin to move about. Sufficient aeration gave the bed the characteristics of a fluid—a pressure head, mobility and zero angle of repose.

These liquid-like characteristics made possible the movement of the catalyst from reactor to regenerator and back with ease, and so the term fluidized was coined. The new system might just as easily have been termed a mobile bed reactor or a turbulent bed reactor. The confusion due to the term fluidized is evident when a layman views the process and expresses surprise that the coal is not molten. For to the layman, fluidized implies a liquid.

Also, there is a commercially successful boiler which utilizes a partially fluidized bed on a moving grate. The differences between the Office of Coal Research process described in this book and the moving grate system are so profound that confusion should not exist. Yet, since each has been termed a "fluidized bed boiler," they are constantly being compared.

The Fluidized State Spectrum: To illustrate better the principles of fluidized bed combustion, Figure 1.1 shows the sequence of and relationship between the various phenomena that comprise the entire fluidized state spectrum as demonstrated by the Wilhelm and Valentine apparatus.

Solids can be fed at a constant rate while the air rate is gradually reduced from its original very high value in (A). Owing to this high air velocity, the particles pass through section 1 out of the apparatus. The void fraction is very high and the pressure drop per unit length of tube is virtually that of an empty tube. As the fluid rate is reduced (B), the particle population increases, the void space decreases slightly and the pressure drop rises.

At this point, the solids will still move concurrently upward with the air. Another small reduction in air rate (C), causes the solids to crowd together and violent slugging occurs, with the solids moving downward. The void space decreases sharply. A condition of minimum voidage is observed in (D) and the solids now have a well defined upper boundary.

The condition represented in (D) is in effect, that experienced in the dense-phase fluidized bed. With a further reduction in air flow, solids inventory in the bed now decreases to reach a limiting value dictated by the rate of addition at port S.

In industrial processes, virtually all the states in the fluidized spectrum find application. The bulk of the applications, however, make use of the dense-phase fluidized bed and combustion applications are no exception.

Similarities Between Fluidized Sand and a Liquid: The term fluidization was obviously invented to describe a certain mode of contacting granular solids with fluids (whether liquid or gaseous) and where the solids also behave with pseudo-fluidic characteristics.

In Figure 1.2a, the sand forms a so-called fixed bed, the sand particles contained in the bed are motionless and supported by contact with each other. In contrast to the fixed bed, a fluidized bed (Figure 1.2c) is a bed in which the individual sand granules are disengaged somewhat from each other and may be readily moved around with the expenditure of much less energy than would be required if the bed were not suspended in the air stream.

In its mobility, the aerated sand column resembles a liquid of high viscosity. If the outside wall is tapped, ripples and waves in the surface of the sand are formed, very much like those observed on liquid surfaces. A small wooden boat placed on top of the bed will pitch and roll as if the sand were a liquid, while a rock will sink to the bottom. Neither rock nor boat will sink in an unfluidized bed.

The depth of solids has also a characteristic hydrostatic head in the fluidized bed. If an opening (Figure 1.2b) is made in the cylinder wall, the fluidized sand will pour out while the unfluidized sand will not. The top profile of a fluidized bed, it will be noted, also tends to maintain a level interface.

FIGURE 1.1: THE WILHELM AND VALENTINE APPARATUS

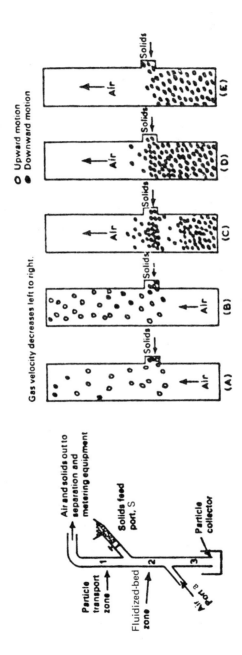

FIGURE 1.2: SIMILARITIES BETWEEN FLUIDIZED SAND AND A LIQUID

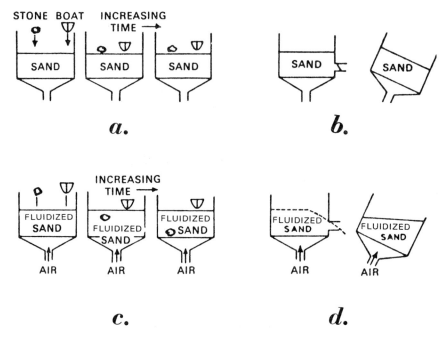

Source: *Technology Ireland,* Dec. 1976

These properties, reminiscent as they are of the properties of liquids, make it appear that the solids bed has been rendered fluid. Hence, the operation of achieving this is termed fluidization.

Description of a Fluidized Bed Boiler

A fluidized bed boiler is defined as a system which meets all of the following criteria:

(a) The system's primary function is the generation of steam. Therefore, the materials of construction, the mode of operation, the arrangement, auxiliary power requirements, etc., are consistent with existing practices and economics in the conventional boiler field.

(b) Fuel is added to and burned within a turbulent bed which has been termed a fluidized bed.

(c) A significant fraction of the heat released by the fuel is extracted by heat transfer surfaces in direct contact with the burning fuel bed.

ASPECTS AND ADVANTAGES OF THE FLUIDIZED BED BOILER

Basic Configuration

Figure 1.3 is a simplified sketch of the essential components of a fluidized bed boiler. A chamber is constructed, preferably using boiler tubes, joined together, as the walls. These are termed "water walls" and are on the order of 18" apart. The steel plate which forms the bottom of the chamber is perforated so that air may pass upward and into the chamber. Into the perforations are placed noz-zles specially designed so that air may pass upward and into the chamber. Noz-zles and plate together are termed the "air distribution grid."

An inert granular material is crushed, passed through a one-eighth inch screen and placed in the chamber to form a static bed about 12 inches deep. The bed material may be any substance which will not burn or melt at temperatures of about 1800°F. It should also not attrite excessively. Coal ash clinkers from a stoker have been found to constitute a suitable starter bed after crushing and screening. Certain natural stones also exhibit the required characteristics. Lime-stone or dolomite makes a suitable bed material and both are good for air pol-lution control.

FIGURE 1.3: SCHEMATIC FLUIDIZED BED BOILER

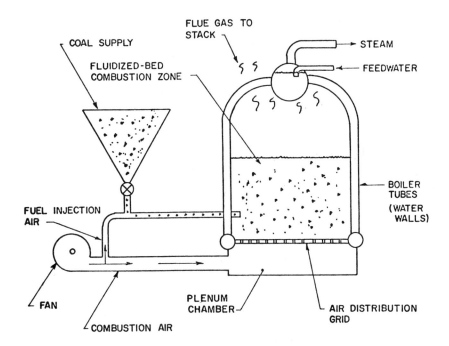

Source: PB 234 343

Beginning Operation

Air from the forced draft fan is introduced into the plenum chamber which is the air chamber beneath the air distribution grid. Because of the design of this grid, the air travels through the nozzles and enters the chamber evenly and with considerable turbulence. As the air flow is increased, the ash granules which form the bed begin to move about and finally become fluidized. A gas burner (not shown in the sketch) is ignited and the flame is directed into the bed. Although the gas flame temperature is above the melting point of the ash particles, the particles do not melt since they move in and then out of the flame zone rapidly. When the temperature of the fluidized bed has risen to about 800°F, some coal is added to the bed and begins to burn. This further increases the temperature of the bed.

As the coal flow is increased and the air flow held constant, the quantity of unreacted oxygen remaining in the flue gas decreases. Efficient combustion is possible when the oxygen content of the flue gas is as low as 1% (5% excess air). The ability to burn coal with excess air requirements equivalent to natural gas flames is one of the advantages of the fluidized bed boiler. When stable combustion is achieved, the start-up gas flame is extinguished.

Performance

To meet the goals of the Office of Coal Research Development program, coal and air flow should be sufficient to release about 1,000,000 Btu/hr/ft^2 of grid surface. If the boiler is to be rail transportable, the plan dimensions can not exceed about 11' x 40'. If 1,000,000 Btu/hr were released for each of the 440 ft^2 and absorbed by steam generating surfaces, then approximately 350,000 pounds per hour of steam may be produced. Coal is therefore fed at the rate of about 80 lb/ft^2/hr. This coal is in the form of roughly crushed particles. In the experimental system, the coal is crushed so as to pass through a one-fourth inch screen. The expense of pulverizing the coal is avoided and this is another of the advantages of the fluidized bed boiler. Neither is it necessary to be concerned with the quantity of fines in the coal feed as in stoker fired boilers.

Because of the mixing action of the turbulent bed, coal particles as small as 325 mesh and as large as 1 inch will be distributed throughout the bed. The fluidized bed boiler will burn any size or type of coal fed regardless of ash, moisture or volatile content, coking or caking characteristics, ash softening characteristics, etc. This is one of the major advantages of the fluidized bed boiler.

The ability to burn any solid fuel stems directly from the nature of fluidized bed combustion. The fuel particles actually form a very small fraction of the bed, in some cases less than 2%. Each fuel particle is therefore almost totally surrounded by an inert refractory material which does not compete for the available oxygen supply but provides an almost infinite supply of ignition energy.

The fluidized bed boiler has some of the properties of a refractory firing chamber, but the "refractory," being the coal ash clinker, or stone starting bed, requires no maintenance and is provided at negligible or no cost. This is another of the advantages of the fluidized bed boiler.

Selection of Dimensions

Referring to Figure 1.3, the water walls which are about 18" apart form the enclosure. This tube bank spacing plays a key role in the operation of the fluidized bed boiler for these dimensions permit operation with uniform bed temperatures and very little excess air.

If no heat were carried away from the bed except via the products of combustion, the bed temperature would rise beyond the melting point of the ash particles and the bed would collapse. That this would be so can be seen from the simple heat balance which follows.

If, instead of boiler tubes, the walls of Figure 1.3 were refractory, the wall loss would be very small. Assume that the combustion air enters at room temperature and, therefore, contributes no heat. The gas loss and the fuel input accordingly must be equal. With low excess air, about 900 pounds of gas are produced in releasing 1,000,000 Btu from the fuel. If 900 pounds of gas are to carry away all of the 1,000,000 Btu, the temperature of the gas (and hence the bed through which the gas had passed) must be about 3700°F. Since such a system does not achieve reasonable temperatures, a sizable wall loss is required.

For a temperature of 1800°F, the 900 pounds of gas would carry away 440,000 Btu and the water walls must absorb the remaining 560,000 Btu. Because of the scrubbing action of the turbulent bed, heat is readily transferred to the water in the tubes.

An overall heat transfer coefficient of about 50 British thermal units per square foot per hour per degree Fahrenheit ($Btu/ft^2hr°F$) has been measured by several methods. If the water in the tubes is at 400°F and the bed at 1800°F, heat flows into the tubes at the rate of 70,000 Btu/ft^2hr. In a pulverized fuel boiler, the comparable figure is 20,000 to 25,000 Btu/ft^2hr. This tripling of the average heat flux from the combustion zone is a major advantage of the fluidized bed boiler.

If heat is released in the bed at the rate of 1,000,000 $Btu/hr/ft^2$ of grid plate area and 560,000 Btu must be absorbed by tubes at the rate of 70,000 $Btu/hr/ft^2$ of tube surface, then for each square foot of grid, there must be 8 square feet of heat transfer surface, i.e., boiler tube surface. In a pulverized fuel boiler, where the furnace exit gas temperatures should also be 1800°F, 24 square feet of exposed boiler tube surface might be required to absorb the same 560,000 Btu.

From this comparison the fluidized bed boiler has an apparent three to one advantage on boiler tube surface when compared to the pulverized fuel boiler furnace. However, the advantage actually is greater and might approach six to one. The water walls in the pulverized fuel boiler receive heat from just one side since they are positioned only on the walls of the furnace. These walls may be 50 feet apart.

In the fluidized bed boiler design indicated schematically by Figure 1.3 repeating elements are placed side by side on 18" centers to provide the desired steam capacity. This ability to distribute effective heat transfer surfaces evenly through the combustion zone is another of the major advantages of the fluidized bed boiler. What actually occurs is far more complex than just described. Other complexities that must be considered are: distribution of heat transfer surfaces; defining the heat transfer surface; preheating the air; and reducing the size of the coal fired boiler. These are described in other sections of the book where appropriate.

The Fluidized Bed Utility Boiler

The principles of the fluidized bed boiler may be applied to boilers designed for large power generating stations. It is in this field, the utility boiler, that this form of steam generation should have the greatest impact on coal sales.

In the large power boiler, the most costly components are the superheaters and reheaters. These components constitute a large fraction of the total tubing and also require, in certain locations, alloy steels. Alloy steels may be as much as 12 times as costly as carbon steel. The cost of superheaters is the area in which the fluidized bed boiler has the greatest impact on utility boiler economics. The factors which make this possible are as follows:

(a)   The high temperature corrosion which has become the leading cause of utility boiler failure should not occur in fluidized bed boilers.

(b)   The safety factors or allowances which must be applied to conventional superheaters and reheaters need not be considered in the fluidized bed boiler.

Although heat flux and, therefore, metal temperature is precisely the same around the circumference of tubes in both conventional and fluidized bed boilers, significant differences do exist. For example, although an actual average metal temperature might allow a lower grade of alloy and/or a thinner tube wall, the necessary safety factors do not permit this.

The most important single factor which makes the superheater designs possible is the relatively low temperature of the fluidized bed. Operating bed temperatures on the order of 1500°F are entirely feasible, and when sulfur retention is important, they are desirable. Superheater surfaces immersed in a fluidized bed at 1500° to 1600°F may be designed without the numerous constraints which apply to superheaters which "see" a 3000°F temperature and have gases above 2000°F passing them.

Fluidized bed boiler principles, when applied to the utility size unit, have a flexibility in form and arrangement that is not available to the packaged boiler design. However, the same economic advantages of factory fabrication should lead to the evolution of a utility boiler built up from rail or barge transportable modules.

Pollution Control and Fluidized Bed Boilers

All boilers can be viewed as chemical reactors in which, among others, the following reactions take place:

(I) $2H_2 + O_2 \rightleftharpoons 2H_2O$

(II) $C + O_2 \rightleftharpoons CO_2$

(III) $S + O_2 \rightleftharpoons SO_2$

However, the fluidized bed boiler more closely approximates what is normally considered a chemical reactor. So, in addition to reactions (I), (II), and (III), operating conditions can be adjusted to promote reaction (IV).

$$\text{(IV)} \qquad CaO + SO_2 + \tfrac{1}{2}O_2 \rightleftharpoons CaSO_4$$

Reaction (IV) may take place in any boiler but may be carried out more completely in a fluidized bed boiler. The ability to adjust the combustion conditions so as to achieve a desired result is a feature which is unique in fluidized bed boilers and constitutes one of its most significant advantages. Because of the relatively low temperatures at which the combustion takes place, the reaction

$$\text{(V)} \qquad N_2 + O_2 \rightleftharpoons 2NO$$

takes place less rapidly and so less completely than in a conventional boiler.

The third significant pollutant from a coal fired boiler is particulate matter. Stoker fired boilers, cyclone furnaces, and wet bottom pulverized fuel boilers have a natural advantage over the fluidized bed boiler in that a significant fraction of the ash comes off as bottom ash and not as fly ash. However, this advantage is not particularly significant, since the particles which leave a fluidized bed boiler are relatively large and easy to collect in a mechanical collector. This is discussed in detail later.

CONCLUSIONS

Fluidized bed combustion of coal is practical. Large fluidized bed boilers can be assembled in factories at relatively low cost. Fluidized bed boilers also offer advantages for air pollution control not found in conventional units. Four important characteristics distinguish fluidized bed combustion from that which takes place in conventional coal fired boilers.

(1) The coal may be burned within a fluidized bed very rapidly and in a limited space. Compared with existing technology, the increase in the volumetric heat release rate is on the order of 10 to 1.

(2) The heat is transferred to the boiler tubes rapidly and, more significantly, uniformly. Compared with existing technology, the improvement in the average heat flux ranges between 3 and 6.

(3) A closely spaced arrangement of the boiler tubes becomes practical. This makes the fluidized bed boiler compact. Compared with existing technology, the ratio "heat transfer surface per unit heat release volume" may be increased by a factor of 10.

(4) A fluidized bed provides a highly reactive zone in which combustion or any other chemical reaction may be carried out very rapidly. Operating temperature reductions of as much as $100°F$ are feasible.

These four characteristics make the following advantageous features possible in fluidized bed boilers:

(1) Cost reduction of 50% from existing technology systems.

(2) Reduction in air pollution, caused by oxides of sulfur or nitrogen, at lower cost than any available alternative.

(3) Increase in thermal efficiency so that more electricity might be obtained from each pound of coal.

(4)   Ability to burn coals having low ash fusion temperatures without creating slagging problems.
(5)   Ability to burn high ash fuels without incurring substantially higher operating costs.

The cost reduction arises not only from the characteristics of fluidized bed combustion, but more importantly from the economies made possible by manufacturing large capacity packaged boilers. A packaged unit is one that is essentially shop fabricated and shipped to the purchaser's site. The advantages of shop fabrication have never previously applied to coal fired boilers above 50,000 pounds per hour capacity. The state of the art in packaging has progressed to the point where it is less costly to purchase and install two packaged units than a single unit requiring assembly in the field.

Design and performance calculations indicate that packaged coal fired fluidized bed boilers of capacities up to 350,000 pounds per hour are possible. Units in this range will generally serve industrial purposes.

While the above applies particularly to boilers of the industrial size, the technical factors apply equally to utility boilers. It is in this area that the significant coal market lies. In the development of an industrial boiler there are definite constraints on size, shape and arrangement, but these constraints need not apply to utility boilers.   Predictions for utility boilers may be based on two facts—one technical and the other economic.

The technical fact is that fluidized bed combustion and heat transfer take place within no more than a five-foot height. This is a fundamental aspect of the process and true for both industrial and utility boilers. There is, therefore, no need to copy the shape and arrangement of pulverized fuel boilers which reflect the fundamental aspects of suspension burning.

The economic fact is the significant savings to be realized by factory fabrication. These savings would accrue to utility boilers too, if they could be fabricated in this manner. Even today the various manufacturers produce numerous shippable subassemblies to minimize the effort in the field. With the advent of the fluidized bed boiler, this concept can be carried to its logical conclusion—a large utility boiler made up of an array of subassemblies each of shippable dimensions.

Since the basic design of the fluidized bed boiler is not significantly influenced by characteristics of the fuel, it is also possible to manufacture the subassemblies in large numbers and then provide an almost "off-the-shelf" utility boiler.

Despite the fluidized bed boiler's potential cost advantages and its ability to burn almost any coal effectively, there may still be some reluctance to accept this new boiler concept—except for its unique air pollution control potential. Because the boiler can be designed to operate at any temperature between about 1450° and 2150°F, sulfur can be captured as it is released by the fuel. Also, it is possible to remove the sulfur in such a way as to permit its recovery and beneficial use. Therefore, recently adopted air pollution regulations will encourage the adoption of fluidized bed boilers and lead eventually to a fossil fueled power economy that is both cleaner and cheaper.

GENERAL STUDIES
OF COMBUSTION PROCESSES

The material in this chapter was excerpted from PB 234 344; PB 260 478; and *Environmental Science and Technology*. For a complete bibliography see p 263.

As stated before, the pressurized fluidized bed combustion of coal can reduce the emission of SO_2 and NO_x from the burning of sulfur-containing coals by using a suitable SO_2 sorbent as the fluidized bed material.

By immersing steam generating surfaces in the fluidized bed, the bed temperature can be maintained at low and uniform temperatures in the vicinity of 800° to 950°C. The lower temperatures decrease steam tube corrosion, allow the use of lower grade coals (since these temperatures are lower than ash slagging temperatures), and also decrease NO_x emissions. Operation at elevated pressures, in the range of 600 to 1,000 kPa, offers further advantages. The hot flue gas from a pressurized system can be expanded through a gas turbine, thereby increasing the power generating efficiency even further. A simple flow diagram illustrating the overall aspects of fluidized bed combustion is shown in Figure 2.1.

A pressurized fluidized bed combustion and regeneration process, as studied by Exxon Research and Engineering Company, is shown in Figure 2.2. Exxon has built two pressurized fluidized bed combustion units to study the combustion and regeneration processes. The smaller of the two units is the batch unit and the larger unit is the miniplant. The descriptions of these two units follow.

MINIPLANT PROGRAM—EQUIPMENT

The Exxon fluidized bed combustion miniplant is shown schematically in Figure 2.3. The miniplant consists of a refractory lined combustor vessel with provisions for continuous feeding of coal and fresh sorbent and continuous withdrawal of spent sorbent. A refractory lined regenerator vessel was built adjacent to the combustor and provides for the continuous transfer of spent sorbent from the combustor to the regenerator and the continuous return of regenerated sorbent to the combustor. The combustor vessel is 9.75 m high, lined to an internal diameter of 31.8 cm, and can burn up to 218 kg/hr of coal.

FIGURE 2.1: FLUIDIZED BED COMBUSTION SYSTEMS

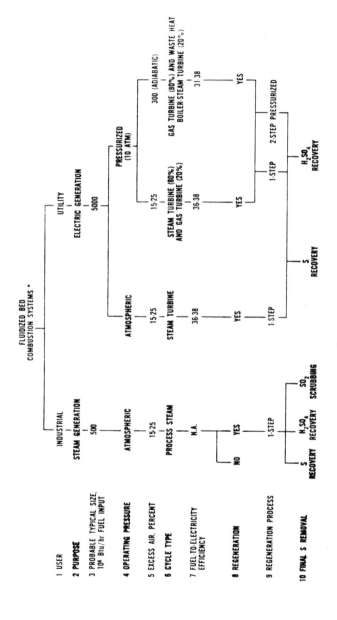

*PROBABLE Ca:S MOLE RATIO IN FEED TO COMBUSTION IS 2:1. COAL SULFUR CONTENT MAY BE 4 PERCENT IN INITIAL SYSTEMS. PROBABLE OPERATING TEMPERATURE OF THE MAIN COMBUSTOR IS 850-950° C.

Source: *Environmental Science & Technology*, March 1977

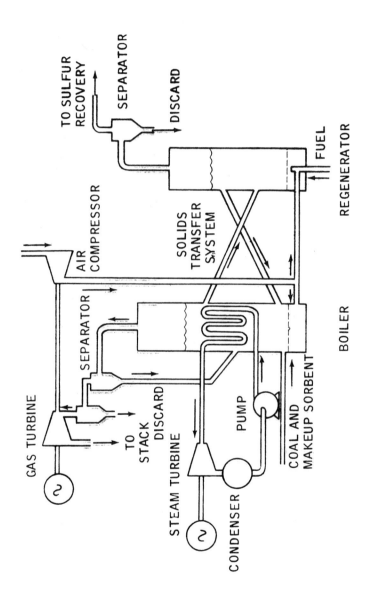

FIGURE 2.2: EXXON PRESSURIZED FLUIDIZED BED COAL COMBUSTION SYSTEM

Source: PB 260 478

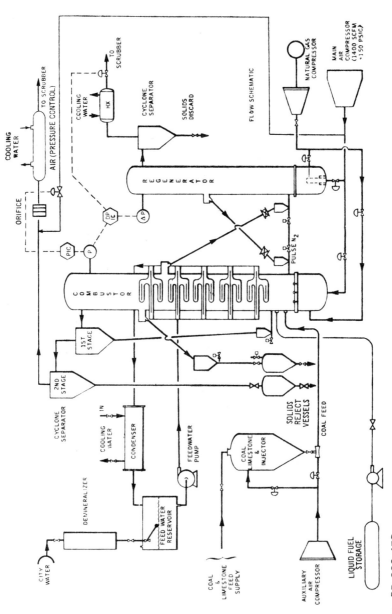

FIGURE 2.3: EXXON FLUIDIZED BED COMBUSTION MINIPLANT

Source: PB 260 478

Cooling coils in the combustor remove the heat of combustion and maintain the bed temperature in the operating range of 800° to 1000°C (1470° to 1830°F). The maximum operating pressure is 1,010 kPa (10 atm); the maximum superficial velocity is 3 m/s (10 ft/sec). The regenerator vessel is 6.7 m high (22 ft), lined to an internal diameter of 21.6 cm (8.5 in) and is capable of operating at temperatures up to 1100°C (2000°F) at 1,000 kPa (10 atm) pressure.

Solids Feeding System

The solids feeding system is illustrated in Figure 2.4. The system provides for uninterrupted solids feed (coal and limestone) from the primary injector to the combustor while allowing intermittent refilling of the primary injector (193 kg operating capacity) when its charge is reduced below 102 kg.

Solids in the primary injector are continuously aerated at a controlled pressure above that in the combustor. They exit the primary injector through a 1.3 cm diameter orifice and are pneumatically conveyed by a controlled stream of dry transport air through an s-shaped 1.3 cm i.d. stainless steel pipe leading into the combustor. A short segment of 1.3 cm i.d. rubber hose is used to connect the injector to the transport line in order not to interfere with operation of the load cells, located under the injector, which are used to monitor solids feed rate. The rate of solids feed is automatically controlled in order to maintain a specific operating temperature within the combustor. This is accomplished through a series of controls involving the pressure differential between the primary injector and combustor, the injector pressure, and the transport air flow rate.

Final entry of solids into the combustor is through a 1.3 cm i.d. probe located 28 cm above the fluidizing grid and horizontally extending about 2.5 cm beyond the reactor wall. The tip of the probe includes ten 0.79 mm diameter holes which surround the solids feed opening. They are used to continuously inject an annular stream of sonic-velocity air to assist penetration of the solids feed into the fluidized bed and alleviate any tendency of the probe tip to become blocked with bed solids.

The remainder of the solids feeding system is involved in generating a specific feed ratio of coal to limestone, and refilling the primary injector in response to an appropriate weight demand signal. Crushed and sized coal and limestone are stored separately in 13.6 tonne and 1.8 tonne capacity storage bins, respectively. From here they are screw-fed at a preselected ratio through a blender into the feed injector (91 kg operating capacity). Transfer from the feed injector to the primary injector is done pneumatically.

Prior to initiation of a refilling operation, the primary injector, feed injector, and pair of solids storage bins remain isolated from each other. When the load cell under the primary injector detects a solids loading of less than 102 kg, 91 kg of solids are automatically transferred from the pressurized feed injector without interrupting feed to the combustor. Refilling is usually completed in about 5 minutes. After refilling, the feed injector is again isolated from the primary injector, vented, and filled with solids from the storage bins. The feed injector is again isolated and repressurized to await repetition of another cycle.

An auxiliary air compressor (rated capacity of 200 scfm at 200 psig [5.6 standard m^3/min at 1,380 kPa gauge]) provides process air for the solids feeding system. Prior to contacting solids, air is dried using a Pall Heatless Regenerative Dryer.

FIGURE 2.4: COAL AND LIMESTONE FEED SYSTEM

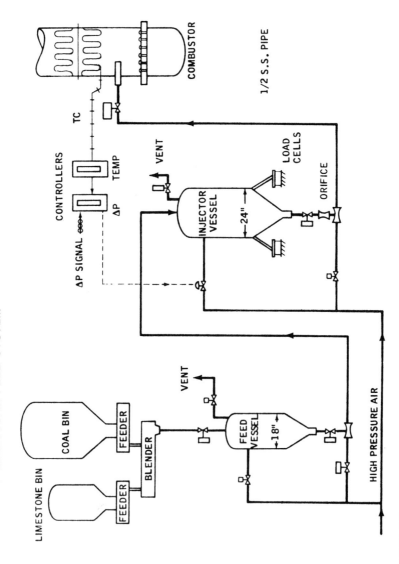

Source: PB 260 478

Fluidized Bed Combustor

The combustor consists of a 61.0 cm i.d. steel shell refractory lined with Grefco No. 75-28 Litecast to an actual internal diameter of 31.8 cm. The 9.75-m high unit is designed in flanged sections and contains various ports to allow for material entry and discharge. Numerous taps are also provided along the length of the combustor to monitor both pressure and temperature.

Pressure measurements are made at five locations: immediately below and above the fluidizing grid; at two points within the expanded bed; and at one location above the bed. Temperatures within the combustor are continuously measured via protected thermocouples at twelve vertical locations, extending from 15.2 cm above the grid to the top of the unit, as well as at one point within the plenum below the grid. Combustion air to the unit is provided by the miniplant main air compressor having a capacity of 1,400 scfm at 150 psig (40 standard m^3/min at 1,030 kPa gauge).

Combustor Startup Burner

The bottom plenum section, where combustion air enters, also houses the burner used for initially preheating the limestone bed during unit startup. Fuel to the burner is provided by a natural gas compressor with a capacity of 20 scfm at 200 psig (0.57 standard m^3 /min at 1,379 kPa gauge). The burner is shown schematically in Figure 2.5.

FIGURE 2.5: COMBUSTOR PREHEAT BURNER

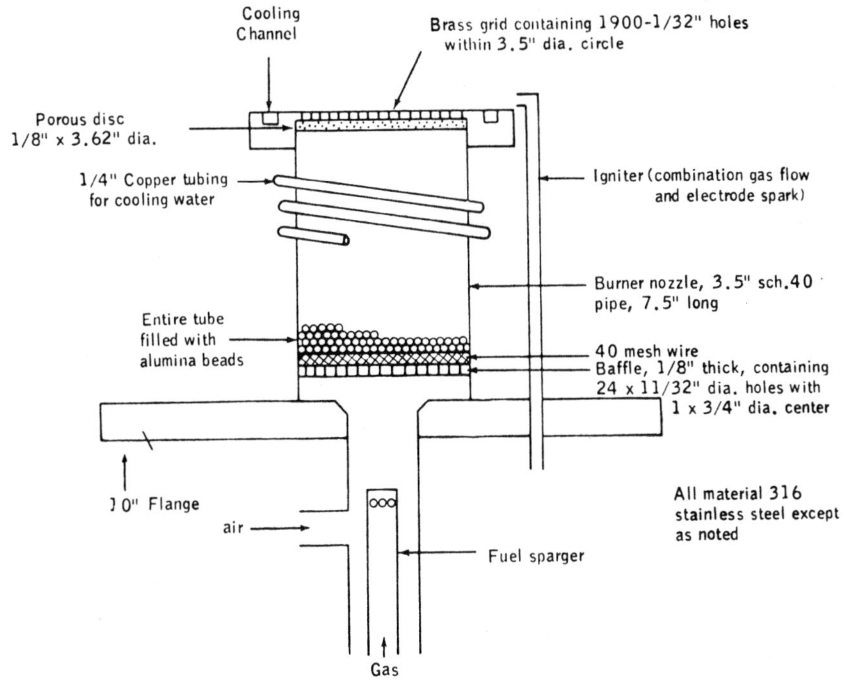

Source: PB 260 478

A well-mixed stream of natural gas and air is ignited as it exits the nozzle using a flame generated by an auxiliary spark-fired pilot ignitor. The burner is water-cooled (both skin and cooling water temperature are continuously monitored) and is designed to avoid flash-back during operation.

Once the fluidized bed temperature reaches 430°C, a liquid fuel (kerosene) system is used to heat the bed to the coal ignition temperature (>650°C). This system is also used to maintain bed temperature above 650°C should coal feeding be interrupted during normal combustor operation. Liquid fuel entry is at a point 15.2 cm above the combustor fluidizing grid. Injection is made through a 1.6 mm i.d. spray nozzle mounted within a 1.6 cm i.d. pipe, resulting in a 0.76 mm annular spacing at the tip. Air flowing through the annulus provides cooling and creates a high velocity stream to insure penetration of the fuel into the bed as well as to avoid clogging of the nozzle.

Combustor Fluidizing Grid

The combustor fluidizing grid, located 71 cm above the combustor bottom, separates the plenum from the main combustor chamber. The grid consists of a 1.3 cm thick x 69 cm diameter stainless steel plate containing 776 equally spaced holes of 0.16 cm diameter within a 31.8 cm circle. The design is such as to give a grid pressure drop of approximately 30% of the total pressure drop across a 2.5 to 3.0 m deep expanded bed. The unit is water-cooled with both the cooling water temperature and metal temperature continuously measured.

Combustor Cooling Coils

Heat removal from the combustor is provided by cooling coils located in discrete vertical zones above the grid. Each coil has a total surface area of 0.55 m² and consists of vertically oriented loops constructed of stainless steel pipe. A coolant distributor plate accommodates each pair of coils with one coil extending 45.7 cm above the plate while the other extends the same distance below. The plates, in turn, are sandwiched between combustor flanges located at 91.5 cm vertical increments beginning 91.5 cm above the fluidizing grid.

Demineralized cooling water is pumped from a storage tank through the coils where it is partially vaporized. After exiting the combustor, the steam/water mixture flows through a condenser prior to return to the storage tank. Flow to each of the coils can be separately monitored and controlled. Cooling water exit temperature from each coil is also routinely recorded. In order to simplify measurement of heat transfer coefficients, it is also possible to increase the water flow to any one of the coils to prevent vaporization during the time of measurement.

Combustor Solids Rejection and Transport

Solids rejection from the combustor is required to maintain a steady-state bed height whenever a mixture of coal and limestone is fed. Such rejection is accomplished through a port located 230 cm above the fluidizing grid. From here, solids flow by gravity through a 15.4 cm i.d. steel pipe, refractory lined to 5.1 cm, into a refractory lined pulse pot. Next, they are pneumatically transported by controlled nitrogen pulses to a pressurized lockhopper from which they are periodically dumped into metal drums.

The full operational mode of the miniplant involves simultaneous operation of both the combustor and regenerator. In such a scheme, limestone would be continuously transferred from the combustor to the regenerator and back again after appropriate regeneration. Solids transfer between both units is to utilize pulse pots similarly to that used for combustor solids rejection. Solids are to exit the combustor through another port located 230 cm above the fluidizing grid and return from the regenerator through a port located at the same height as the coal feeding point (28 cm above the fluidizing grid).

Combustor Cyclones

Flue gas and entrained solids (fly ash and limestone) exit the top of the combustor through a 45.7 cm i.d. steel pipe, refractory lined to 18.4 cm i.d., and enter a two-stage cyclone system. Both cyclones are refractory lined and designed to operate at combustor temperatures and pressures.

Solids (primarily limestone) separated by the first-stage cyclone drop through a 20.3 cm i.d. steel dip leg, refractory lined to 10.2 cm i.d., and enter a refractory lined pulse pot. From here they are pneumatically conveyed back to the combustor using controlled nitrogen pulses and enter at a point 66 cm above the fluidizing grid. Solids (primarily fly ash) escaping the primary cyclone pass through a 30.3 cm i.d. steel pipe, refractory lined to 14.5 cm i.d., and enter the more efficient second-stage cyclone, where a finer distribution of solids is captured. Collected solids then pass through a 20.3 cm i.d. dip leg, refractory lined to 10.2 cm i.d., and enter a pressurized lockhopper from which they are periodically dumped into metal drums. A Ducon granular bed filter with appropriate modifications placed after the second-stage cyclone is to capture solids that manage to escape the cyclone separation system.

Pressure Control and Flue Gas Discharge System

The technique used to control combustor pressure consists in dropping the system pressure across an appropriately sized ceramic coated nozzle located in the flue gas exit line. Back pressure is controlled by regulating the flow of a secondary air stream to the nozzle inlet.

After exiting the second-stage cyclone, flue gas passes through 20.3 cm i.d. steel pipe refractory lined to 10.2 cm i.d., and is expanded through a converging nozzle. The 2.5 cm thick nozzle consists of a conical-shaped entrance section with an initial diameter of 3.5 cm which converges at an angle of 30° over a distance of 1.0 cm to a 2.3 cm diameter cylindrical throat of 1.5 cm length. The nozzle is constructed of mild steel and is mounted in a carbon steel flange. The exposed surfaces of the nozzle and flange are covered by a 0.1 mm thick flame-sprayed coat of chromium carbide. In addition, a subcoat of nickel aluminide flame sprayed to the base metal surfaces provides a sound substrate for adherence of chromium carbide.

The remainder of the combustor pressure control system is located 0.6 m upstream of the nozzle and consists of a secondary source of high pressure air which is metered through a 5.0 cm Kamyr ball valve equipped with a pneumatic actuator and positioner. By superimposing a secondary flow of air onto the primary flow of flue gas through the nozzle, control of combustor pressure is maintained at the desired level which has been typically 910 kPa.

Downstream of the nozzle, flue gas passes through a 2.5 m 316 stainless steel pipe, water jacketed by a 4 inch carbon steel pipe, and enters a scrubber for final cleanup before venting to the atmosphere. During transit from the upstream side of the nozzle to the scrubber entrance, flue gas temperature is reduced from approximately 820° to 260°C, with 35 to 50% of the reduction due to expansion through the orifice and injection of the low temperature secondary air stream.

Flue Gas Sampling and Analytical System

Flue gas is sampled at a point about 7 m downstream of the second-stage cyclone exit (about 4.5 m upstream of the pressure control nozzle). A schematic of the sampling system is shown in Figure 2.6. Other sampling systems are shown in a later chapter.

The system is designed to produce a solids-free, dry stream of flue gas at approximately ambient temperature and atmospheric pressure, whose gas composition is unaltered (except for water content) from that of the original flue gas. In order to allow for cross-checking and backup of the miniplant's main analytical instrument train, the system includes provisions for removing samples for wet chemistry determinations and also direct routing of a stream to the batch fluidized bed combustion unit analytical instruments.

FIGURE 2.6: FLUE GAS SAMPLING SYSTEM

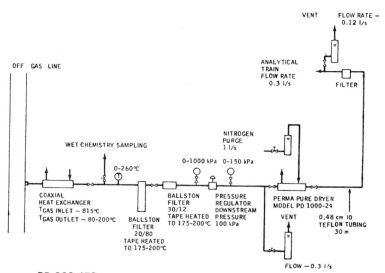

Source: PB 260 478

Process Monitoring and Data Generation System

Three multichannel recorders record output from various measuring instruments. In addition, at one-minute intervals, the same output is recorded by a data logger system with printer and a magnetic tape recorder with the electronic interface between the two.

Approximately 60 pieces of data are logged with three-quarters involving temperature measurement while the rest deal with pressure and material flow rate.
Signals from the data logger appear as digital output on printed paper tape and
are also stored on magnetic tape. The magnetic tape, containing about 3,600
items of data per hour of run time, is fed to a computer which converts the logger output to flow rates, pressures, etc. with the proper dimensions. The data
are then averaged and standard deviations calculated over preselected time intervals (usually 10 minutes). Other quantities are also calculated. These include
average bed temperature (based on four thermocouple readings covering the 15
to 114 cm interval above the fluidizing grid), superficial gas velocity, and excess
air.

Combustor Safety and Alarm System

A process alarm system warns of impending operational problems. Two general
alarm categories exist. The first, dealing with less critical situations, alerts the
operator of the problem so that appropriate corrective action can be taken. The
second class of more critical alarms results in the immediate or time-delayed shutdown of the complete system or specific subsystems.

Fluidized Bed Regenerator System

The regenerator reactor, designed for operation at 1100°C and pressures up to
1,010 kPa, consists of a 45.7 cm i.d. steel shell, refractory lined with 75-28
Grefco Litecast to an internal diameter of 21.6 cm. Numerous taps are provided
along its 6.66 m overall height to monitor both temperature and pressure, while
appropriately located ports allow for material entry and exit. Discussion of various regeneration systems are described in a later chapter.

Miniplant Support Structure

The flanged steel beam support structure for the miniplant is 12.9 m high,
9.0 m wide and 3.9 m deep. Three platforms at 2.4 m, 5.7 m, and 9.0 m (with
connecting stairways) are provided for servicing the miniplant. The structure
rests 18.4 cm off the ground on concrete support pillars. The reactors are supported by the structure at the first platform. Thermal expansion joints are provided at various locations to accommodate thermal expansion of the reactors in
the vertical direction. A five-ton bridge hoist, mounted on top of the support
structure, is used for assembly and disassembly operations.

MATERIALS USED IN MINIPLANT STUDY

Coal

The coal used was a high volatile bituminous coal obtained from Consolidation
Coal Company. It was obtained from the Arkwright Mine in West Virginia
and from the Champion preparation plant in Pennsylvania. Both coals are
Pittsburgh No. 8 seam coals and have similar analyses. Grinding and sizing was
done by Penn-Rillton Company. Essentially all of the coal was less than 2,380 μm
(No. 8 U.S. mesh) in size. Actual size distributions used during the course of
the program are given in Figure 2.7. Two size distributions were used. The original batch contained all the fines. Batches prepared for subsequent runs had

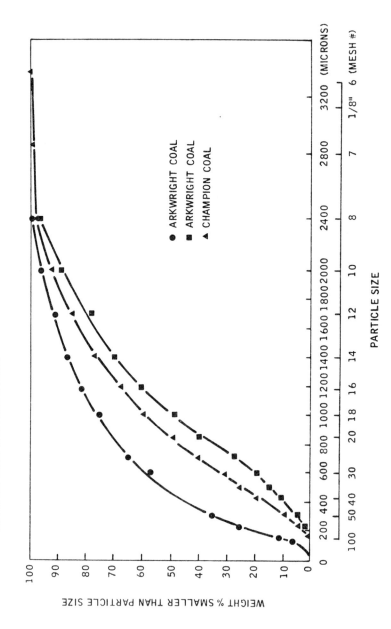

FIGURE 2.7: COAL PARTICLE SIZE DISTRIBUTION

Source: PB 260 478

most of the fines smaller than 40 mesh removed. This gave a distribution more closely resembling that expected to be used in commercial FBC units. Composition analyses are shown in Table 2.1.

TABLE 2.1: MINIPLANT COAL ANALYSES

Component	Arkwright	Arkwright	Champion
	. . . . Weight Percent.		
Moisture	1.0	0.9	2.2
Ash	8.1	7.4	8.8
Total carbon	76.5	77.1	78.1
Hydrogen	5.3	5.1	5.1
Sulfur	2.6	2.5	2.2
Nitrogen	1.5	1.1	1.6
Oxygen (by difference)	5.0	6.0	4.1
Chlorine	0.1	0.1	0.1
Higher heat value (Btu/lb)	14,100	13,700	13,600

Source: PB 260 478

Limestone

The only sorbent used for the miniplant runs was uncalcined limestone obtained from Grove Lime Company (Stephen City, Virginia) designated as Grove No. 1359. This material, in its calcined form, contained 97.0 wt % CaO, 1.2 wt % MgO, 1.1 wt % SiO_2, 0.3 wt % Al_2O_3, and 0.2 wt % Fe_2O_3. The uncalcined limestone feed was screened to give a distribution with a minimum of 90% between 2,380 μm (No. 8 U.S. mesh) and 841 μm (No. 20 U.S. mesh). Actual particle size distribution of the limestone bed during the course of a run would be shifted into the smaller particle size range due to attrition.

MINIPLANT PROCEDURES

Combustor Startup

Prior to initiating a run, a detailed checkout procedure is followed to insure that the system is ready for operation. All runs are begun with an initial bed of limestone in the combustor. This consists of either a fresh charge of uncalcined stone or the bed from the previous run.

The first operation of startup involves preheating the limestone bed using natural gas and then by kerosene. Prior to ignition of natural gas, an air flow of about 350 scfm (9.9 standard m^3/min), or about half that used at normal operating conditions, is fed through the burner while combustor pressure is raised to 280 kPa gauge. Once ignition of the natural gas occurs, this procedure maximizes incoming gas temperature under conditions which allow good natural gas combustion and adequate bed fluidization. Ignition begins by simultaneously feeding 20 scfm (0.57 standard m^3/min) of natural gas through the burner while activating an ignition electrode. Because of the limited capacity of the gas compressor, natural gas is used only to heat the bed to a temperature of about 430°C, sufficient to insure self-ignition of kerosene. This generally requires 20 to 30 minutes. At this point, kerosene is injected into the lower portion of the bed. When rising temperatures indicate ignition of liquid fuel, natural gas feed is discontinued to insure sufficient air for complete combustion of kerosene. Approximately 10 to 15 minutes are required to raise the bed temperature to 650°C, which is sufficient to achieve self-ignition of coal.

Coal, usually mixed with limestone, is then fed to the combustor from the primary injector. A steady stream of 60 scfm (1.7 standard m^3/min) of transport air is used to convey coal into the combustor. Actual rate of coal injection is determined by the pressure differential between the injector and combustor. The rate is initially set at an appropriate value based on past experience under similar operating conditions.

Once ignition of coal is verified by rapidly rising temperatures, kerosene flow is stopped. At this time, the main combustion air feed line to the plenum is opened allowing most of the air to bypass the burner, and both combustion air flow rate and combustor pressure are rapidly increased to their designated operating values. Flow of water to each cooling coil is adjusted to maintain steam/water exiting temperatures of 138° to 150°C. Once the desired bed temperature has been reached, it is held approximately constant by the automatic coal feed rate control system.

Combustor Shutdown

A run is terminated by first discontinuing coal feed which results in a rapid decrease in bed temperature. As temperature falls, fluidizing air flow rate and combustor pressure are decreased stepwise. When temperature falls below 90°C, which generally requires 10 to 15 minutes, air flow is halted and the combustor is depressurized. At this time, remaining systems, including cooling water flows, are shut down and a nitrogen purge is introduced into the combustor to prevent condensation of moisture.

BATCH COMBUSTOR STUDIES—EQUIPMENT

For purposes of comparison, a brief description of the batch combustor unit is presented here. A schematic diagram of the Exxon batch fluidized bed combustion unit is shown in Figure 2.8. The combustor is equipped with a continuous coal feeding system. The sorbent is added batch-wise. The primary components of the unit are the coal feeding system, the fluidized bed combustor, and the gas handling and analytical equipment.

Coal Feeding System

Figure 2.9 shows a Petrocarb Model 16-1 ABC injector. The main features are a conical-bottom tank that holds solids to be fed, and an orifice and mixing tee assembly that mixes solids with carrier gas. Solids in the tank are aerated by a controlled stream of air at a selected pressure. Aerated solids flow through the orifice at the bottom of the tank into the mixing tee assembly and are picked up by a controlled stream of carrier gas (air). Solids are pneumatically conveyed through a transport line into the combustor. The feed rate of solids is controlled by pressure in the feed tank, carrier or injection air flow rate, and pressure differential between the feed tank and combustor.

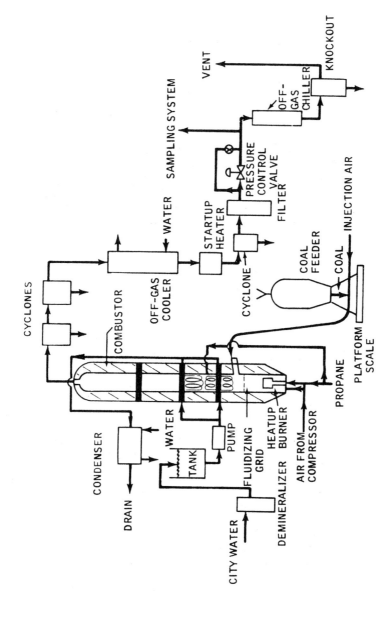

FIGURE 2.8: BATCH FLUIDIZED BED COAL COMBUSTION UNIT

FIGURE 2.9: PETROCARB COAL INJECTOR

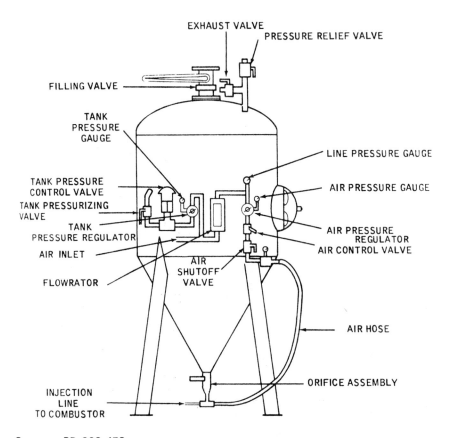

Source: PB 260 478

In this case the Petrocarb solids feeder was modified to feed ground coal (–16 mesh) at rates of 3 to 13 kg/hr into the combustor against pressure of up to approximately 1,000 kPa. The feeder, as it is supplied by Petrocarb, can handle only much higher feed rates. The diameter of the orifice was reduced to 0.71 cm and the Petrocarb injection hose was replaced with a 0.58 cm diameter i.d. by 6.10 m long stainless steel tube.

In order to make the feeder work satisfactorily with the batch combustor, the feeder to combustor pressure differential had to be held constant. This was accomplished with automatic controls which maintained the pressure in the feed tank above the pressure in the combustor by the desired amount. The entire feeding assembly was mounted on a platform scale which measured the coal feed rates.

Pope, Evans and Robbins, Inc. have tested many coal feed systems. Under a recent contract seven basic systems were examined with varying results. The optimized system employs 45° downward inclined feed tubes, each being a singly supplied split tube (double port) serving the six-foot length of the fluidized bed module (FBM) boiler and a three-foot bed width of the proposed boiler. Each coal take-off thus serves 18 square feet of bed with above-bed excess O_2 variation held to ±0.5%. This represents a significant improvement over the double throw downflow feeder (±3% variation) used in early experimental work. Tubes are of straight-in configuration and can be removed and replaced during boiler operation.

Evenness of distribution is enhanced by use of ¾ inch top-sized coal, but a classification system must then be developed to limit the amount of over-sized bed particles. A decision on the use of ¼ inch top-sized coal without classification versus ¾ inch top-sized coal with classification has to be made.

Fluidized Bed Combustor

A schematic diagram of the batch fluidized bed coal combustor is given in Figure 2.10. The vessel was constructed from four sections of 25 cm (10 inch) diameter standard wall carbon steel pipe and refractory lined with Grefco Litecast No. 7528 to an inside diameter of 11.4 cm. The height of the vessel, above the fluidizing grid, was about 4.9 m. Below the grid was a 61 cm long burner section lined with Grefco Bubbalite. The fluidizing grid, which was made of stainless steel, had eighty 0.16 cm diameter holes to distribute the fluidized air and was water cooled.

A propane burner, located at the bottom of the burner section, was used to preheat the unit to above the self-ignition temperature of propane (505°C). At this point propane was added directly into the bed just above the fluidizing grid to raise the temperature of the bed above the self-ignition temperature of coal. The maximum operating temperature and pressure of the batch combustor were approximately 1000°C and 1,000 kPa respectively.

The combustor had three vertical cooling coils made of 0.95 cm diameter (o.d.) stainless steel tubing. They extended from 27 to 141 cm above the fluidizing grid and each had a surface area of 0.060 m². Each coil had its own rotameter and control valve and the water flow into each coil and the temperature of the steam issuing from the coils was easily controlled. Thermocouples were located 15 cm apart in the lower section of the combustor and 30 cm apart in the upper section.

Sorbent was loaded into the combustor through a charging port located in the upper section. Solids could be removed through a port in the lower section or, alternatively, transferred directly to the adjacent batch regenerator by blowing them through a 5 cm diameter pipe supplied for this purpose.

Coal entered through a "sonic" coal injection probe which was connected to the end of the 6.10 m x 0.58 cm diameter (i.d.) coal injection tube. The stream of flowing coal was surrounded by seven sonic air jets. The primary intent of the high velocity air jets was to improve coal feeding by clearing a path through the bed of fluidized solids in the combustor. Air flow through the annulus of the probe helped to cool the probe. This flow was about 1.89 dm³/s and compares to an air flow of about 3.30 to 3.78 dm³/s used to transport the coal.

FIGURE 2.10: COMBUSTOR VESSEL

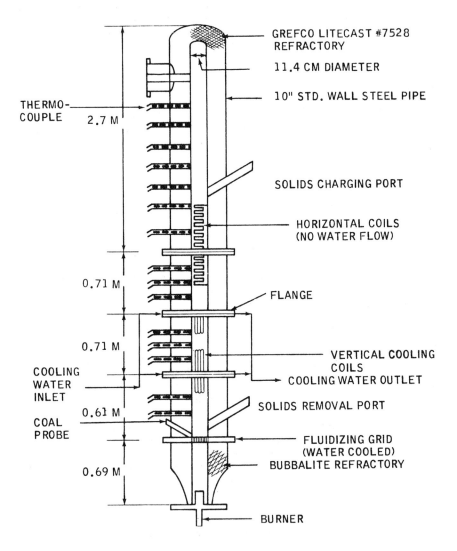

Source: PB 260 478

Gas Handling and Analytical Equipment

Flow of air and fuel into the combustor, and system pressure, were under auto-matic control. Gases leaving the combustor first passed through two cyclones, which removed entrained sorbent and fly ash. An off-gas cooler, which followed the cyclones, reduced the temperature of the off-gas to the desired level. The off-gas then entered a coil of 2.5 cm diameter stainless tubing which was elec-trically heated during startup to raise the temperature of the gases above the dew point. A 3.81 cm diameter Aerotec cyclone, following the heater, was used to remove particulates during startup of the combustor, when water vapor condensation in the first two cyclones caused them to operate at reduced efficiency. Fine particulates were removed with a Pall filter, located upstream of the back-pressure control valve. This filter had a mean pore size of 165 microns and an area of 0.37 m².

Before being vented, the off-gas entered a chiller and knockout to remove moisture so that the water content of the gas could be determined. A small portion of the off-gas was diverted after the back-pressure control valve and sent to the gas sampling system. A schematic of the sampling system is shown in Figure 2.11.

FIGURE 2.11: FLUE GAS SAMPLING SYSTEM

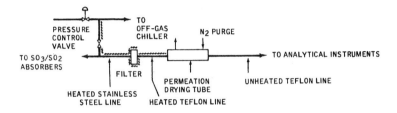

Source: PB 260 478

The sample gas passed through a small length of heated stainless tubing before entering a filter for final particulate cleanup. The gas then went through a 1.8 m section of heated Teflon line before passing through a permeation drying tube for moisture removal. Downstream of the dryer the line was unheated Teflon tubing to the analytical instruments.

MATERIALS USED IN BATCH COMBUSTOR STUDIES

Coal

Three different coals were used in the batch fluidized bed coal combustion program. The majority of the runs were made using a high volatile bituminous coal from Consolidation Coal Company's Arkwright mine in West Virginia. It was ground to –16 mesh. A measured particle size distribution corresponded very closely to the specified distribution given in Table 2.2.

TABLE 2.2: SPECIFIED COAL PARTICLE SIZE DISTRIBUTION

U.S. mesh size	10	20	30	40	100	200	pan
Friction on screen, wt %	0	4.5	15.5	14	35.5	12.5	18

Source: PB 260 478

Runs were also made with a low sulfur Western subbituminous coal and a high sulfur Illinois bituminous coal. The Illinois coal was provided by Argonne National Laboratory. A proximate and an ultimate analysis were made on each of the coals and the results are presented in Table 2.3.

TABLE 2.3: COMPOSITION OF COALS USED IN BATCH FLUIDIZED BED COAL COMBUSTION PROGRAM

	. . . .Coal Source* (weight percent). . .		
	1	2	3
Proximate			
Moisture	1.00	6.0	2.2
Ash	8.11	12.7	7.9
Volatiles	36.86	40.5	44.0
Fixed carbon	54.03	40.8	45.9
Ultimate			
Moisture	1.00	6.0	--
Ash	8.11	12.7	7.9 (dry)
Total carbon	76.26	62.9	68.0
Hydrogen	5.30	4.5	5.0
Sulfur	2.66	4.1	0.7
Nitrogen	1.49	1.2	0.9
Chlorine	0.07	0.0	0.0
Oxygen**	5.11	8.5	17.9

*(1) Arkwright—Consolidation Coal Co.; higher heat value = 14,045 Btu/lb.
(2) Illinois—Peabody Coal Co.; higher heat value = 11,300 Btu/lb.
(3) Wyoming—Carter Oil Co.; higher heat value = 11,850 Btu/lb (dry).

**By difference.

Source: PB 260 478

Sorbents

Grove limestone (BCR No. 1359) and Tymochtee dolomite were the primary sorbents used in the experimental studies. The stones and their properties are given in Table 2.4. The particle size of these materials was generally in the 8 x 25 mesh range. Baker dolomite and Pfizer dolomite (BCR No. 1337) were also tried but, because of their high attrition rates, their use was discontinued.

TABLE 2.4: PROPERTIES OF SORBENTS USED IN BATCH FLUIDIZED BED COAL COMBUSTION PROGRAM

Designation	Quarry Source	Stone Type	. . . . .Chemical Analysis (wt %).				
			CaO	MgO	SiO_2	Al_2O_3	Fe_2O_3
1359	Grove Lime Co. (Stephen City, Va.)	Limestone	97.0	1.2	1.1	0.3	0.2
Tymochtee	C.F. Duff & Sons (Huntsville, Ohio)	Dolomite	53.8	38.7	5.3	0.9	1.2

Source: PB 260 478

BATCH COMBUSTOR PROCEDURES

Experimental Operation

Operation of the batch fluidized bed combustor can be divided into four phases: startup, ignition and preheating, coal feeding, and shutdown. Startup consisted of those activities preliminary to ignition of the propane burner. These activities included checking equipment to make sure it was ready for a run, checkout of the analyzer calibration, charging sorbent, turning on electrical circuits and the air compressor, turning on all cooling water systems (fluidizing grid, burner, steam coils, condenser) and purge air systems (pressure taps, sight-glasses, ΔP cells).

To ignite the propane burner, air and fuel flows were set and the ignition electrode was activated. Safety devices shut down all flows if ignition was not obtained within ten seconds or if a flame-out occurred afterwards. A safety interlock prevented startup for 3 minutes after an automatic shutdown to assure adequate purging of the combustor. Subsequent to ignition, combustor air flow and pressure were adjusted to the values desired for making the run. All gas flows and pressure were controlled automatically. After the bed temperature reached the ignition temperature of propane, propane flow to the bed was initiated to heat the bed up to the coal ignition temperature.

Preparation of the coal feed system for a run consisted of setting the flow of injection air and activating and adjusting the coal feeder-to-combustor ΔP control system. Coal injection could be started only after the temperature in the combustor was high enough for self-ignition of the coal to occur. Propane flow was stopped automatically at the same time that feeding of coal was started. An automatic safety circuit would shut down coal injection if the combustor temperature dropped too low to ensure combustion of the coal or if the feeder-to-combustor ΔP dropped below a preset minimum (about 1 psi).

Data on the weight of the coal feeder versus time was taken so that the feed rate of coal could be determined. Another method of estimating the feed rate was to observe the oxygen concentration in the off-gas from the combustor. A rapid rise in oxygen concentration was usually the quickest way of determining that a problem was developing with the coal feeding system.

Temperatures in the combustor could be controlled by regulating the amount of coal burned. The feed rate of coal could be adjusted by changing the flow of injection air or coal feeder-to-combustor ΔP.

To shut down the combustor routinely, the coal feed valve was shut, fluidizing air was stopped, and nitrogen purge was started to preserve the solids. Flow of injection air was kept on for several minutes so that the coal feed line could be cleared of coal. All water flows were reduced. Solids could be discharged from the reactor (by blowing them out of a port located just above the fluidizing grid) after the bed cooled overnight.

COMBUSTION CONFIGURATIONS
AND CYCLE DESIGNS

The material in this chapter was excerpted from:

CONF-760402-4
FE-1514-42
PB 231 977
PB 237 028
PB 246 116

For a complete bibliography see p 263.

Fluidized bed combustion technology can be applied with many different configurations and can be utilized in many different power cycles. Several pressurized fluidized bed combustion boiler and adiabatic combustor designs and alternative pressurized fluidized bed combustion power plant cycles are presented here. Operation, control and performance of these systems are evaluated.

BASE SYSTEMS

Pressurized Fluidized Bed Boilers

A preliminary design for a commercial-scale, pressurized fluidized bed boiler power plant has been prepared by Westinghouse. Their results follow. A 320 MW plant required four modules of 3.7 meter (12 foot) diameter, and a 635 MW plant required four modules of 5.2 meter (17 foot) diameter. Each pressure vessel housed four separate fluidized beds of cross section 1.5 by 2.1 meters (5 by 7 feet). The bottom bed was for preevaporation, the top bed for reheating and the other two beds for superheating.

Evaporation of water was accomplished in the water walls which enclosed the beds. The design was characterized by vertical modular components, horizontal steam tubes immersed in the bed, depths of 3.3 to 4.3 meters (11 to 14 feet), gas velocities of 1.8 to 2.7 m/sec (6 to 9 ft/sec), pressure of 1,013 kPa (10 atm), and limestone/dolomite for sulfur removal.

The effect of the change in fluidized bed operating conditions and design parameters on economics and performance was analyzed (1) and found to be essentially invariant with the projected design basis. The effects of steam pressure and gas-turbine pressure ratio were also evaluated.

A pressurized fluid bed combustion boiler combined-cycle power plant using state-of-the-art power generation equipment with steam conditions of 16,548 kPa/538°C/538°C (2,400 psi/1000°F/1000°F), gas-turbine pressure ratio of 10:1, and gas-turbine temperature of 871°C (1600°F) was the preferred design. The calculated heat rate was about 9,040 kJ/kWh (8,570 Btu/kWh).

British Coal Utilization Research Association (BCURA) developed a 140 MW pressurized fluidized bed boiler design for a combined-cycle power plant (2). The design was characterized by a horizontal 4.3 meter (14 foot) diameter, 30.5 meter (100 foot) long pressure shell, horizontal steam tubes immersed in the bed, gas velocity of 0.6 meter (2 feet) per second, bed depth of 1.2 meters (4 feet), pressure of 810 kPa (8 atm), and no provision for sulfur removal. Subsequent to this initial design, a 70 MW design was proposed.

The BCURA design operates at a fluidizing velocity of 0.8 m/sec (2.5 ft/sec), and at a pressure of 1,611 kPa (16 atm). A pressurized fluid bed combustor pilot plant has been successfully operated at BCURA to obtain process data (3)(4)(5).

Pressurized Fluidized Bed Adiabatic Combustors

The requirement for an internal heat transfer surface in fluidized bed combustion can be eliminated by increasing the excess air to a bed operating at 927°C (1700°F) to approximately 300%. An adiabatic coal-fired fluidized bed combustor is applicable to gas-turbine or combined gas-turbine/steam-turbine cycles. Combustion Power Company has operated a pilot plant of a gas turbine with a fluidized bed combustor on prepared solid waste. The unit was modified to test coal (6). The pilot plant combustor is 2.7 meters (9 feet) in diameter, operating at 414 kPa (60 psi) with gas velocities up to 2.1 meters (7 feet) per second and 0.6 meter (2 foot) bed depths.

The product gas goes to a Rustin TA 1500 gas turbine. Westinghouse has carried out conceptual designs and performance and economic studies on adiabatic fluidized bed combustion systems for coal-fired combined-cycle power plants. Two combustor design concepts were studied, a single fluid bed module and a stacked fluid bed module. The fluidized bed combustors were designed to operate at gas velocities of about 1.8 meters (6 feet) per second, bed depths of 1.8 to 2.1 meters (6 to 7 feet) and bed temperature of 954°C (1750°F). With a turbine inlet temperature of 927°C (1700°F), the calculated plant heat rate is about 9,600 kJ/kWh (9,100 Btu/kWh).

ALTERNATIVE PRESSURIZED FLUIDIZED BED COMBUSTION POWER CYCLES

Advanced Steam Conditions

In the late 1950s, the trend toward higher power plant efficiency with little regard for economics culminated in the building of the Eddystone Unit No. 1, with throttle steam conditions of 34,475 kPa/649°C/566°C/566°C (5,000 psi/1200°F/1050°F/1050°F) (7). Because of serious operating problems with this unit, and with the design for 31,028 kPa (4,500 psi), subsequent supercritical units have been limited to pressures of 24,133 kPa (3,500 psi). More recently, the trend has been away from supercritical cycles because they were relatively unreliable and uneconomical, and unable to attain their predicted performance in practice.

During the early 1960s, a number of steam plants were constructed with super-heat temperatures of 593°C (1100°F). Operating experience with these plants showed that there was no economic advantage in using steam temperatures above 538°C (1000°F), and since that time most new units have been designed for 538°C (1000°F superheat and reheat.

A description of the typical coal-fired power plant of the early 1970s is as follows:

(1) 800 MWe, 3,600 rpm, tandem compound, four 31-inch low-pressure ends
(2) Seven feed heaters plus gland, generator, and oil coolers, 5° approach
(3) 16,548 kPa/538°C/538°C (2,400 psi/1000°F/1000°F)
(4) Condensing at 8.4 kPa (2.5 in Hg), natural draft wet towers
(5) Turbine cycle heat efficiency of 44% [8,271 kJ/kWh (7,840 Btu per kWh)]
(6) Unit net heat rate of 9,622 kJ/kWh (9,120 Btu/kWh)
(7) Pulverized coal fired
(8) Consumptive water use 1.7 kg/kWh/50,000 m³ per day (3.75 lb/kWh/11,000,000 gpd)

During the initial phases of the preliminary design of the high-pressure fluidized bed boiler for utility applications, a parametric study was made to compare the estimated performance and energy costs of high-pressure fluidized bed boilers operating at 16,548 (2,400) and 24,133 kPa (3,500 psi). For both pressure levels the superheat and reheat temperatures were 538°C (1000°F). The study indicated that there was no economic advantage in using supercritical steam conditions in the high-pressure fluidized bed boiler with superheat and reheat temperatures of 538°C (1000°F). Experience with conventional pulverized-coal plants has shown that there is little economic advantage to using 24,133 kPa (3,500 psi) steam pressure instead of 16,548 kPa (2,400 psi).

There is reason to think that the hot-side corrosion problems in the fluidized bed boiler will be less severe than in the conventional pulverized-coal boilers and, therefore, that superheat and reheat temperatures greater than 538°C (1000°F) would be technically feasible. In view of this a series of cycle performance calculations was made to see how much the heat rate of a plant with high-pressure fluidized bed boilers could be improved with higher steam superheat and reheat temperatures (and correspondingly higher pressures). The results of these calculations for a gas turbine with a pressure ratio of 10, a turbine inlet temperature of 871°C (1600°F), and an air equivalence ratio of 1.1 are shown in Table 3.1.

TABLE 3.1: PLANT HEAT RATE AS A FUNCTION OF STEAM CONDITIONS

Throttle pressure, kPa,(psi)	Superheat temperature, °C(°F)	Reheat temperature, °C(°F)	No. of heaters	Heat rate kJ/kWh (Btu/kWh)	Ratio of heat rate to base case
16,548 (2400)	538 (1000)	538 (1000)	7	9040 (8570)	1.000
24,133 (3500)	538 (1000)	538 (1000)	7	8900 (8440)	0.985
31,028 (4500)	649 (1200)	649 (1200)	8	8190 (7770)	0.907
34,475 (5000)	760 (1400)	760 (1400)	8	7840 (7430)	0.868

Source: PB 246 116

Cost estimates of high-pressure fluidized bed boiler systems with advanced steam conditions gave the results shown in Table 3.2 for a gas turbine with a pressure ratio of 10:1, a turbine inlet temperature of 871°C (1600°F), and an air equivalence of 1.1. The increased cost of the high-pressure steam turbine for steam temperatures higher than 538°C (1000°F) with corresponding pressures more than offsets the effect of the improved heat rate on the cost of energy.

TABLE 3.2: ENERGY COST AS A FUNCTION OF STEAM CONDITIONS—
PLANT HEAT RATE AS A FUNCTION OF PRESSURE RATIO

Throttle pressure, kPa (psi)	Superheat temperature, °C(°F)	Reheat temperature, °C(°F)	Specific Cost, $/kw	Cost of energy, $/kw
24,133 (3500)	538 (1000)	538 (1000)	374	21.28
31,028 (4500)	649 (1200)	649 (1200)	474	23.70
34,475 (5000)	960 (1400)	760 (1400)	534	25.19

Source: PB 246 116

Oxygen-Blown System

An evaluation of an oxygen-blown atmospheric-pressure fluidized bed boiler led to the conclusion that the high cost of oxygen prohibits the use of an oxygen-blown atmospheric-pressure fluidized bed combustion system for economical steam or power generation. The bare cost of 95% oxygen, using depreciation rates allowable for utilities, and direct costs only for fuel and labor, is $6.81/Mg. The additional charges incurred by an industrial producer of oxygen such as Linde or Air Products add at least $7 or $8 more per Mg of oxygen and give over-the-fence costs of about $14.

Since there would be no significant cost reduction associated with an oxygen-fired plant, an increase in energy costs of up to 10 mills/kWh would result.

Low Temperature Cleanup Techniques for High Pressure Fluidized Bed Boilers

A study was made to determine the performance penalty which accompanies reduced temperature techniques for removing particulates from the combustion products of a high pressure fluidized bed boiler. Two alternatives were investigated.

(1) cooling the combustion products by the use of a convection-type boiler and removing particulates by cyclone separators, tornado separators, electrostatic precipitators, or combinations thereof, at intermediate temperatures; and

(2) cooling the combustion products with a recuperator followed by a scrubber-cooler.

The arrangement for the high pressure fluidized bed boiler system with intermediate temperature particulate removal is shown in Figure 3.1. The convection-type boiler for cooling the products of combustion to a temperature well below the bed operating temperature is in series for the combustion products and in parallel for the working fluid with the fluidized bed boiler. This permits the temperature of the combustion products to be reduced from 870°C (1600°F) to the gas-turbine idle temperature which is in the range of 482° to 538°C (900° to 1000°F).

FIGURE 3.1: HIGH PRESSURE FLUIDIZED BED BOILER SYSTEM WITH INTERMEDIATE TEMPERATURE PARTICULATE REMOVAL

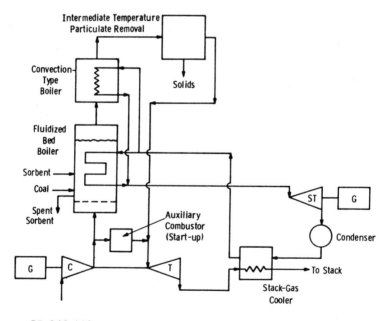

Source: PB 246 116

Analysis of the performance of this system shows that the plant capacity decreases about 1.8%, and the plant heat rate increases about 158 kJ/kWh (150 Btu per kilowatt hour) for each 55.5°C (100°F) drop in temperature.

The arrangement of the high pressure fluidized bed boiler with low temperature particulate removal is shown in Figure 3.2. The hot combustion products are passed through one stage of cyclone separation to recover the larger fraction of the char particles which are elutriated from the bed so that the carbon losses will be reduced to a minimum value. The effluent from the char separator will retain the finer ash particles with relatively low carbon content.

This stream will be cooled to a temperature in the range of from 93° to 204°C (200° to 400°F), depending on the recuperator effectiveness and the temperature of the cold products of combustion out of the scrubber-cooler. In the scrubber-cooler the products are evaporatively cooled to the saturation line and then further cooled along the saturation line until the water vapor content of the gas mixture is equal to the initial value of the fluidized bed boiler.

FIGURE 3.2: PRESSURIZED FLUIDIZED BED COMBUSTION POWER PLANT WITH COLD CLEANUP OF COMBUSTION PRODUCTS

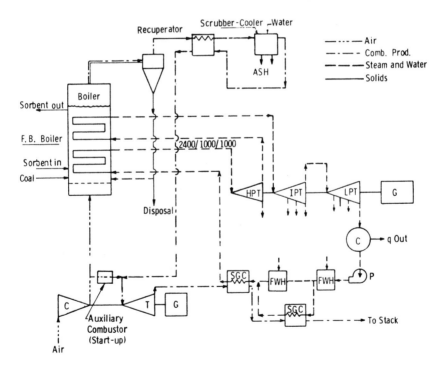

Source: PB 246 116

This stream will be cooled to a temperature in the range of from 93° to 204°C

(200° to 400°F), depending on the recuperator effectiveness and the temperature of the cold products of combustion out of the scrubber-cooler. In the scrubber-cooler the products are evaporatively cooled to the saturation line and then further cooled along the saturation line until the water vapor content of the gas mixture is equal to the initial value out of the fluidized bed boiler. Performance calculations to determine the effects of air equivalence ratio, boiler outlet temperature, system pressure ratio, and recuperator effectiveness indicate:

(1)   The plant heat rate increases about six percentage points for each ten percentage point decrease in recuperator effectiveness.
(2)   Plant heat is a rather weak function of the pressure ratio, and the optimum pressure ratio is about 8.5:1.
(3)   The plant heat rate varies only about 1% over the range of boiler outlet temperatures from 704° to 926°C (1300° to 1700°F).
(4)   Plant heat rate is a rather strong function of the air equivalence ratio, with the heat rate increasing as the air equivalence ratio increases.

Economics for the options considered for the PFBB include the use of hot pressurized electrostatic precipitators for gas cleanup ahead of the turbine, and heat exchange of the boiler off-gases down to water-scrubbing temperatures prior to injection into the turbine.

Gas Turbine Cycle with Indirect Air Cooled Fluidized Bed Combustor

One variation of the adiabatic combustor concept is the utilization of the excess air required to maintain the bed temperature in air-cooled heat transfer surface in the bed. This open cycle gas turbine with a pressurized fluidized bed combustor/air heater has been described by Harboe (8). This permits part of the air (about 70%) from the compressor to by-pass the combustor gas flow, pass through the heat transfer surface immersed in the fluidized bed, where it approaches the bed temperature, and be mixed with the products of combustion after they have been cleaned of particulates.

This alternative is less demanding of the particulate control system than the pressurized boiler and adiabatic combustor base designs. The cycle performance will be slightly lower than the adiabatic combustor system for a given bed temperature, since the excess air will only approach the bed temperature and result in a slightly lower gas turbine inlet temperature. The capital cost, compared with the adiabatic combustor, will depend on a trade-off between an increase in the combustor cost due to the air cooled heat transfer surface and a decrease in the particulate removal system cost due to reduced volumetric gas flow. The plant reliability may be increased if the particulate loading to the gas turbine can be significantly reduced over that achieved with the adiabatic combustor. This may result in a lower development risk.

Balance Pressure Reheater Cycle

Progress in the art of steam power generation depends on innovation as well as on an analysis and extension of existing concepts. An improved, combined cycle, called "superreheat with vapor phase recuperation," is shown in Figure 3.3. This cycle consists of the steps shown on the following page.

(1)   High subcritical or supercritical vapor generation in a boiler which
        has been fired with gas turbine exhaust and fuel which need not
        be "clean"
(2)   Superheating the vapor to the 538°C (1000°F) range in an ortho-
        dox superheater
(3)   Turbine expansion to reheat pressure near the saturation line
(4)   Reheating in a steam-steam recuperator to the 538°C (1000°F)
        range
(5)   Reheat in "Balanced Pressure Superreheater" to 816°C (1500°F)
        or higher
(6)   Turbine expansion of steam from 816°C (1500°F) to First Law
        balance point
(7)   Expanded hot reheat steam recuperating Item 4 above to pinch
        point
(8)   Pinch point steam expanding to condenser.

This apparatus includes a reheat exchange and a high pressure ratio gas turbine
in "Velox" arrangement. The reheat exchanger is a carbon steel, refractory lined
drum containing a radiant convection steam reheat exchanger made of thin walled,
high alloy tubes. Gas-side pressure is approximately equated to steam-side pressure. Clean fuel is fired to the drum and again at the drum exit. Reheated gas
is expanded in the gas turbine. Gas turbine exhaust is cooled against feed heating, steam generation, and (possibly) steam generation at cold reheat pressure.
The gas turbine may beneficially consist of a free gas generator and a reheated
power turbine.

This unusual arrangement is meant to maximize the temperature of heat addition
by means which do not require material breakthroughs in terms of cost, corrosion resistance, and stress-rupture properties. The superreheat turbine is at reasonable pressure for thermal stress minimization, and its steam path may be cooled
with saturated (or near) cold reheat steam. The main features are:

(1)   Use of gas turbine cycle compressed air to permit a very high
        temperature, unstressed steam reheater
(2)   Use of steam-steam recuperator to suppress the addition of
        low temperature heat in the steam reheater
(3)   Use of gas turbine reject heat for feed heating to maximize
        the steam flow in the high temperature reheater
(4)   Optimization of gas turbine compressor intercooler pressure
        to emphasize either (a) the minimum heat rate by raising
        the compressor outlet temperature or (b) the maximum gas
        turbine net power by reducing the compressor work.
(5)   Minimization of steam extraction by maximizing the gas
        turbine/boiler exhaust feed heating. This step results in a
        power split which leans toward the gas turbine shaft. Optionally,
        boiler exhaust heat may be used for additional steam genera-
        tion at cold reheat pressure.

The cycle chosen for a rough test of these principles used 24,133 kPa/538°C/816°C
(3,500 psi/1000°F/1500°F) steam coupled to a 34/1 pressure ratio, 941°C (1725°F)
reheated gas turbine. A unit heat rate of approximately 6,858 kJ/kWh (6,500 Btu
per kWh) appears to be possible for gas firing. This relationship implies about
7,913 kJ/kWh (7,500 Btu per kWh).

FIGURE 3.3: RECUPERATIVE SUPERCHARGED REHEATER CYCLE

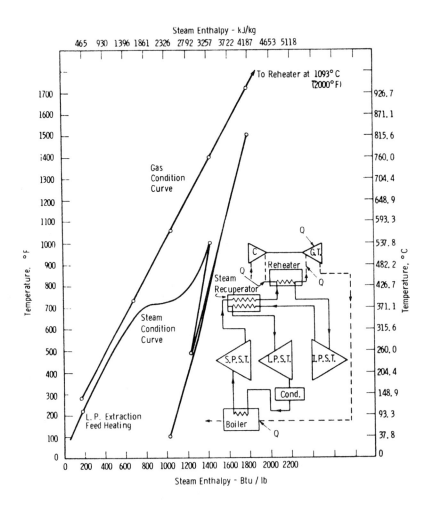

Source: PB 246 116

The superreheat cycle with vapor phase recuperation should be good for application of high pressure fluidized bed boiler technology to an advanced steam system.

Figure 3.4 shows the superreheat with vapor phase recuperation cycle modified to operate with high pressure fluidized bed combustion. There are two high pressure fluidized bed combustors. The first generates and superheats steam at supercritical pressure and operates at an intermediate pressure level. The second fluidized bed combustor is used to reheat the steam up to a temperature of about 816°C (1500°F) with the boiler pressure level approximately equal to that of the reheat steam pressure. The bed temperature and, consequently, the gas turbine inlet temperatures would most likely be equal.

**FIGURE 3.4: SUPERREHEAT WITH VAPOR PHASE RECUPERATIVE CYCLE
WITH HIGH PRESSURE FLUIDIZED BED BOILERS ADDED**

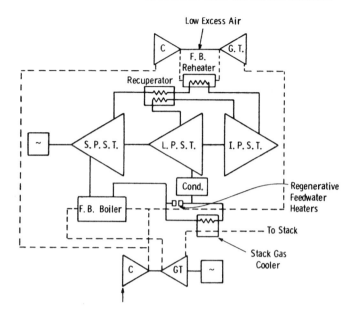

Source: PB 246 116

Liquid-Metal Topping Cycles with Fluidized Bed Combustion

Investigations of liquid-metal topping cycles coupled with pressurized fluidized
bed fired heaters (see Figure 3.5) have been made by Oak Ridge National Laboratory (9) and by General Electric under joint NASA/OCR funding (10). The
conclusions of these studies were that overall plant efficiencies of 50 to 60%
were attainable with this system.

An evaluation of this system concluded that maximum overall plant efficiency
would be in the order of 48% and that the cost of energy would be about 20%
greater than that for the pressurized fluidized bed boiler steam plant with
24,133 kPa/538°C/538°C (3,500 psi/1000°F/1000°F) steam conditions.

Closed-Cycle Gas Turbine Applications

An evaluation of closed-cycle gas turbine systems in which fluidized bed fired
heaters were applied was performed. Figure 3.6 shows a pressurized fired heater
with fluidized bed combustion. When applied to the helium, closed-cycle, gas-
turbine system with steam bottoming shown on Figure 3.7, the calculated plant
heat rate is about 8,750 kW and the estimated capital cost for a plant with construction starting in mid-1974 is about $635/kW. This gives an estimated energy cost
which is approximately 33% greater than that for a pressurized fluidized bed
boiler with steam conditions of 24,133 kPa/538°C/538°C (3,500 psi/1000°F/1000°F).

FIGURE 3.5: SCHEMATIC FOR PRESSURIZED FURNACE POTASSIUM TOPPING CYCLE

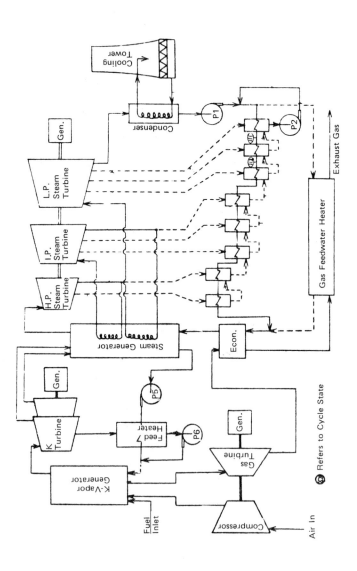

Source: PB 246 116

FIGURE 3.6: PRESSURIZED FIRED HEATER SUBSYSTEM

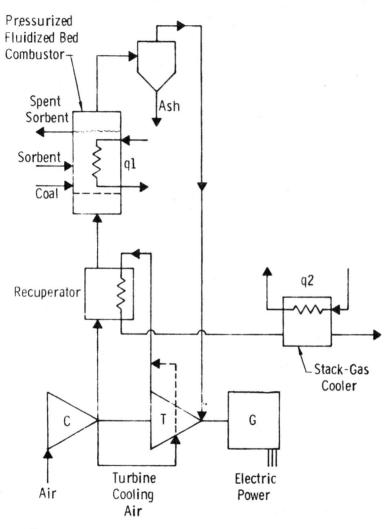

Source: PB 246 116

FIGURE 3.7: CLOSED-CYCLE COMBINED FLOW DIAGRAM

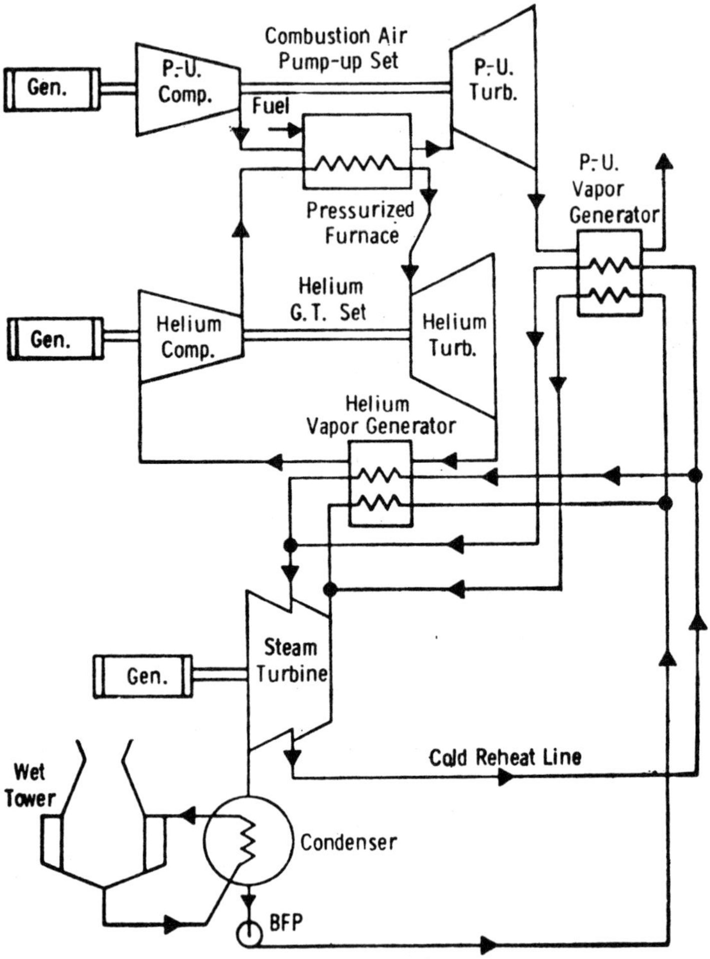

Source: PB 246 116

Technique for Increasing Turbine Inlet Temperature in High Pressure Fluidized Bed Boilers

Turbine inlet temperatures for industrial and commercial gas turbines have historically increased about 14°C (25°F) per year over the past two decades. Current levels are 1038° to 1093°C (1900° to 2000°F) for intermediate load units and 1093° to 1149°C (2000° to 2100°F) for peaking units. These levels significantly exceed the value of 871° to 927°C (1600° to 1700°F) imposed on the gas turbine in the high pressure fluidized bed boilers by the sulfur sorbent. The maximum allowable bed temperature for effective, in-bed desulfurization is thought to be in the range of 954° to 1010°C (1750° to 1850°F), and the turbine inlet temperature is estimated to about 83°C (150°F) lower than the bed temperature.

Cycle calculations were made to determine the effect of turbine temperature on the performance of the high pressure fluidized bed boiler. These calculations indicate that the heat rate of this system improves about 0.5% for every 56°C (100°F) increase in turbine inlet temperature. One possible technique for obtaining turbine inlet temperatures which exceed the limits imposed by in-bed desulfurization is shown in Figure 3.8. Here char which is elutriated from the primary beds is used as a feedstock for a gasifier. The fuel gas produced in the gasifier is burned in a separate combustor to give state-of-the-art gas-turbine inlet temperatures.

The separate gas-turbine combustor for burning the fuel gas from the gasifier would probably be a conventional, integrated type, which has a definite design and cost advantage over an external combustor, especially for high turbine inlet temperatures.

FIGURE 3.8: PRESSURIZED FLUIDIZED BED COMBUSTION POWER PLANT

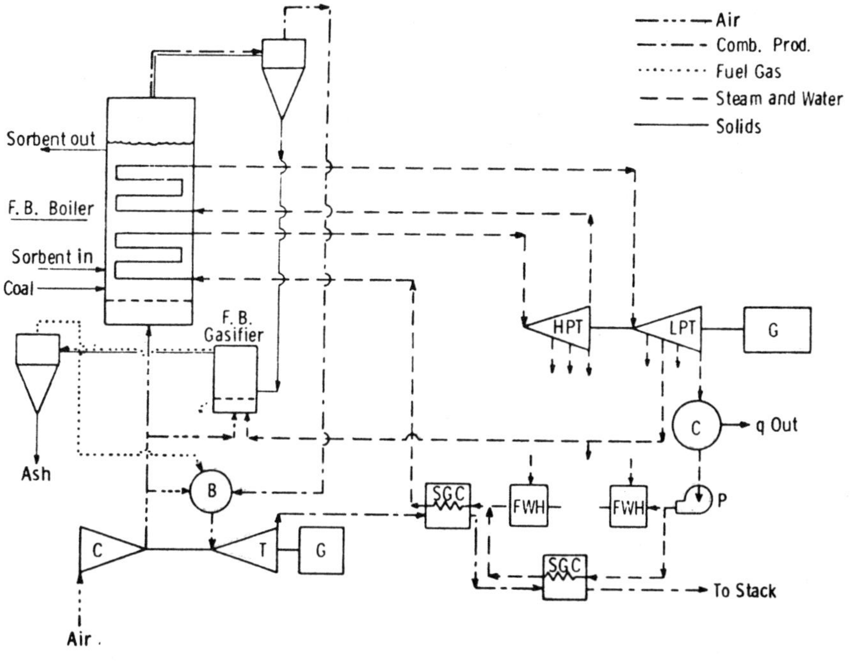

Source: PB 246 116

Modular Integrated Utility System

Oak Ridge National Laboratory (ORNL) has made an evaluation of small coal burning gas turbines with fluidized bed combustion. This modular integrated utility system would use a small on-site utility plant located adjacent to a new housing development to provide electricity, hot water for building heating, domestic hot water, and chilled water for air conditioning. Solid waste from the housing complex would be used as a supplemental fuel in the energy system. These fundamental gas-turbine cycle configurations were considered in this study:

    (1)   Direct-fired open cycle
    (2)   Exhaust-fired open cycle
    (3)   Indirect-fired closed cycle

An indirect-fired closed cycle with an air preheater was selected as the preferred system for this application because of its superior part-load performance. The distribution of input energy in the preferred system was 30% converted into electricity and 60% into steam at 120°C (250°F) and 202.6 kPa (2 atm) or hot water at 65°C (150°F). It was estimated that the overall installed cost of the coal-fired fluidized bed combustion system and gas turbine would be $350 to $400/kWe for production units.

Closed Cycle Gas Turbines: A flow sheet for a typical closed cycle gas turbine system coupled to a fluidized bed coal combustion system is shown in Figure 3.9. For good cycle efficiency a recuperator is employed to recover part of the heat from the hot gas leaving the turbine and use it for heating the gas flowing from the compressor to the heater in the fluidized bed. In addition, a regenerator is employed to recover heat from the hot combustion gas and preheat the combustion air stream fed to the fluidized bed.

Further, waste heat recovery units are employed in the gas stream flowing from the recuperator back to the compressor inlet. In the case of the system of Figure 3.9 that heat would be employed for domestic hot water and building heating in the winter and for absorption air conditioning in the summer. Inasmuch as the system was designed for use with a housing complex, the heat from the lower temperature stage of the heat recovery unit can be used for heating domestic hot water.

Inasmuch as the electrical load demand in a housing complex varies in a somewhat different fashion throughout the day from the heat load demand, it is advantageous to make use of hot water storage tanks so that the power plant can be operated to deliver the electrical power required plus enough heat to maintain an adequate reserve of hot water in the hot water storage tanks.

In very cold weather the recuperator of the gas turbine can be partially bypassed to increase the amount of heat available from the production of a given amount of electricity. This arrangement proved to be eminently satisfactory in operation in the municipal total energy system of the city of Munich, Germany; the cost of the hot water storage tanks can be written off by fuel savings in less than two years. A similar system can be designed to yield process heat for industrial power applications, in which case it may prove desirable to eliminate the recuperator if the amount of process heat or the temperature at which it is required favor such an arrangement.

For operation between the same temperature limits there is no difference in the thermodynamic cycle efficiency between air and helium as working fluids in the cycle, but the size of the heat exchangers can be reduced substantially by using helium rather than air (11).

FIGURE 3.9: FLOW SHEET FOR A CLOSED CYCLE GAS TURBINE
COUPLED TO A FLUIDIZED BED COAL COMBUSTION SYSTEM
FOR TOTAL ENERGY SYSTEMS FOR HOUSING COMPLEXES

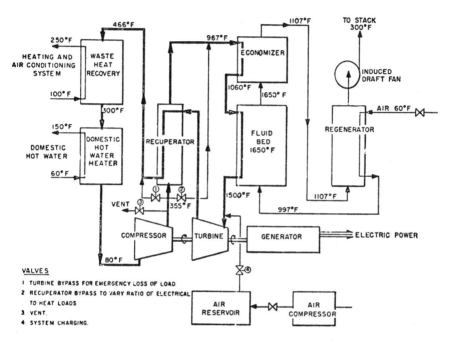

Source: CONF-760402-4

However, the savings in cost stemming from the reduction in size are partially
offset by the increase in cost associated with the most stringent requirements
for leak tightness.

In closed cycle gas turbine systems the power output is controlled by varying
the pressure in the closed cycle; the temperatures throughout the system are substantially constant irrespective of the power. Further, the thermal efficiency of
the plant is essentially independent of the power output down to around 25% of
full design power at which point frictional, pumping, and heat losses become an
increasing fraction of the ideal output.

A fluidized bed furnace designed for use with the system of Figure 3.9 is shown
in Figure 3.10. The air stream from the compressor flows into the upper of the
two cylindrical manifolds shown, upward through a tube bank forming an economizer region where heat is removed from the hot combustion gases leaving the
combustion chamber, then the tube bank loops downward to form the furnace
walls.

From the base of the economizer the tubes loop through the bed, and finally
enter the outlet manifold at the bottom (12). The combustion air enters at the
top of the bell jar surrounding the furnace, flows downward through an outer

annulus just inside the outer shell, and enters the air plenum chamber under the fluidized bed. The combustion gases from the fluidized bed flow upward through the bed and the plenum chamber above it, vertically downward through the economizer region, and then vertically upward in counter flow through an annulus just inside that for the incoming combustion air.

FIGURE 3.10: FLUIDIZED BED COAL COMBUSTION SYSTEM DESIGNED TO SERVE AS HEATER FOR A CLOSED CYCLE GAS TURBINE

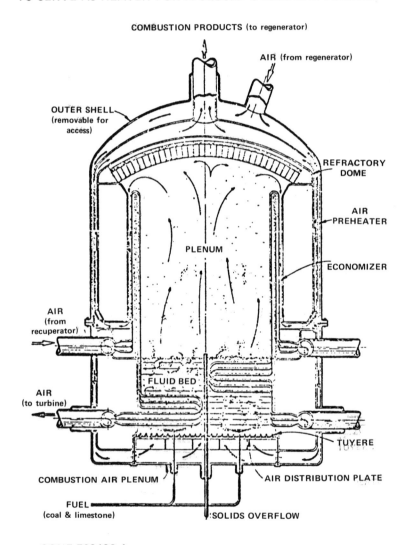

Source: CONF-760402-4

With this arrangement, the outer shell and the plenum chamber structure underneath the combustion chamber are all held at a temperature close to 1000°F (540°C), thus minimizing difficulties with thermal stress and differential thermal expansion (12).

MULTISTAGE GASIFICATION SYSTEMS FOR ELECTRIC POWER GENERATION

Westinghouse System

The development of a new kind of power plant to meet today's need for more economical generation of electric power from caking, sulfur-laden domestic coal in an environmentally acceptable manner is being studied. Analytical, experimental and design work is being performed to demonstrate that the route to future coal-fired power generation of the desired kind lies in perfection of techniques for converting domestic steam coals into suitable gas turbine fuels.

Described here is the Westinghouse fundamental gasification process. As Figure 3.11 shows, the complete system comprises three process units, a preheater, a recirculating bed devolatilizer-desulfurizer and a fluidized bed gasifier-combustor.

FIGURE 3.11: MULTISTAGE FLUIDIZED BED GASIFICATION PROCESS

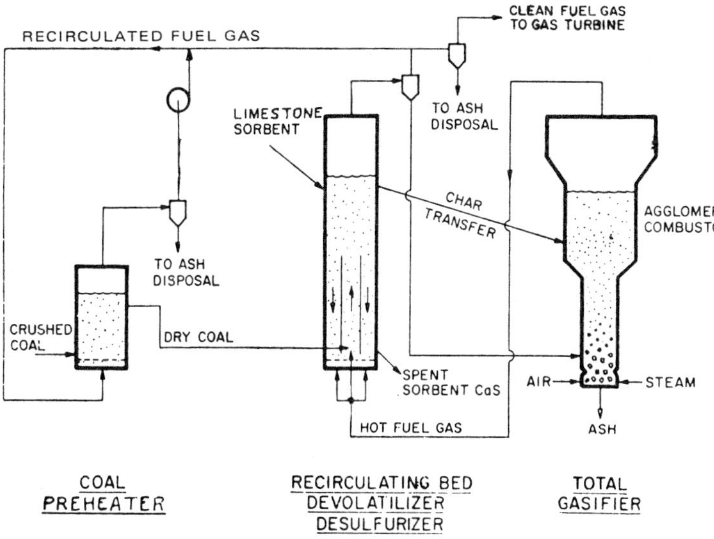

Source: FE-1514-42

Crushed coal is dried in a fluidized bed and then transported to the devolatilizer-desulfurizer unit. Here the devolatilization, desulfurization, and partial hydrogasification functions are combined in a single recirculating fluidized bed reactor. Dried coal is introduced through a central draft tube. Inside this tube, the raw feed coal and large quantities of recycled solids, char and/or sorbent are carried upward by gases flowing from the total gasifier. The recycle solids needed to dilute the feed coal and to temper the hot inlet gases descend in an annular down-comer, a fluidized bed surrounding the draft tube. These recirculating solids, flowing at rates up to 100 times the coal feed rate, effectively prevent or control agglomeration of the feed coal as it devolatilizes and passes through its plastic and sticky phase upon initial heating.

A dense dry char is collected in the fluidized bed at the top of the draft tube. The sorbent is added to this bed in order to remove sulfur, which is present as hydrogen sulfide in the fuel gases. Spent (sulfided) sorbent is withdrawn from the reactor after stripping out the char either in the transfer line or in a separator of special design. Char is withdrawn from the top section of the bed. Heat is primarily supplied to this unit from the high temperature fuel gas produced in the total gasifier. Additional heat is transported to the devolatilizer by solids carry-over in the gases from the total gasifier.

The final gasification of the low-sulfur char is conducted in a fluidized bed with a lower leg which serves as a combustor. In this section, char obtained from the devolatilizer-desulfurizer is burned with air at 2100°F to provide the gasification heat. Steam is injected into this same area. Heat is transported from the com-bustor to the gasifier both by combustion gases flowing upward and by fines which escape upward and are caught and recycled back to the space between the combustor and gasifier.

The ash from fines combustion agglomerates at this temperature on the down-ward-moving ash from the char and segregates in the lower bed leg for removal. Gasification occurs in the upper section of the bed at 1800° to 2000°F, with the sensible heats of both gas and expelled fines providing the heat requirements for the devolatilizer-desulfurizer.

Combustion Power System (CPU-400)

Combustion Power operates two fluidized bed combustor systems. The smaller, termed the "Model Combustor," develops operating parameters and conditions for the larger, "The Process Development Unit," which is designated the CPU-400. The Model Combustor made several runs at varying conditions of coal/dolomite ratios, bed temperatures and bed velocities. An extended run at 1600°F bed tem-perature, Ca/S mol ratio of 1.5, 7 ft/sec bed velocity, Illinois No. 6 coal with 4% sulfur, and 135% excess air provided average emissions of 0.76 lb/10^6 Btu of SO_2 and 0.68 lb/10^6 Btu of NO_2 (reported as NO_x), which compare with EPA stan-dards of 1.0 lb of SO_2 and 0.7 lb of NO_x.

The CPU-400 consists of a pressurized fluidized bed combustor, three inertial separator stages, 1-MW Ruston Model 1500 TA gas turbine and system instru-mentation and compute controls. Subsequent runs on the CPU-400 configura-tion established that all components were operable in the combustion of coal and that attention has to be directed to removal of particulate matter prior to turbine input.

Combustor and Gas Preparation Subsystem: The fluidized bed combustor is con-tained within a vertically oriented cylindrical pressure shell with dished heads. The outside diameter is 9.5 feet and overall height is 23.5 feet. Three layers of insulation protect the three-eighth inch carbon steel pressure shell cylindrical sections from high combustion zone temperatures. A wear-resistant firebrick inner liner is backed up by a liner of insulating brick. These refractory layers are separated from the shell by a thin layer of packed ceramic fiber insulation designed to isolate the shell from stresses induced by differential thermal expansion. Insulation in the top dome is provided by a castable refractory held in place by standard hangers.

The fluidized bed is supported by a flat carbon steel plate welded to the pressure shell and covered by two layers of castable refractories that provide insulation and wear resistance. Penetrating this assembly are 161 two-inch pipes capped with wire mesh air diffusers. Other penetrations from the air plenum chamber beneath the plate permit bed temperature and pressure measurements.

The circular cross section fluidized bed has an area of 40 square feet and is designed to operate with a superficial velocity in the 5 to 7 feet per second range. A nominal 2-foot bed (unfluidized state) is used together with a 12-foot freeboard (unfluidized bed surface to exhaust duct center line).

Penetrations through the pressure shell and refractory insulation into the fluidized bed provide for two solid waste feedpoints. Two feedpipes bolted to outer shell bosses extend into the bed. Solid waste is fed into the bed along the length of these pipes via a slanted cut on the bottom side. The design and positioning of these pipes is based on earlier tests where oxygen concentration measurements established the dimensional characteristics of combustion zones. The result is a configuration which, in low pressure testing, has demonstrated very satisfactory operation with respect to geysering due to feedpipe air flow, minimization of local fuel-rich zones, and reduction of heat release above the bed.

Other bed penetrations provide for possible removal of excess bed material and for six oil guns to permit fluidization combustion of auxiliary diesel oil. This normally unused auxiliary fuel, available primarily as a developmental tool for backup service in maintaining or establishing desired test conditions, is mixed with air and carried through the inner of two concentric pipes. The outer pipe of each gun carries cooling air.

Initial bed heating is accomplished by hot products of combustion from an oil burner located in the top dome of the combustor. This downward firing burner forces hot gases through the bed in a back heating mode that is capable of heating the bed from ambient conditions to 1100°F in 90 minutes. This bed temperature, being above the autoignition temperature of either solid waste or diesel oil, is an appropriate initial condition for successful fluidized combustion.

In the low pressure configuration, fluidizing air is supplied to the combustor's air plenum by a 125 horsepower positive displacement blower which can deliver up to 7,000 scfm at 3 psig. The blower also supplies fuel transport air in a parallel path to the fluidized bed. By appropriate valving, the same blower is used to drive the back heating mode.

The top cylindrical section is removable and contains the exhaust port, instrumentation, and observation ports. Exhaust into the first particulate removal stage is carried by a double-walled pipe with 26-inch carbon steel outer wall and 20-inch type 310 stainless steel inner liner. The annular space is packed with ceramic fiber insulation.

SYSTEM ECONOMICS

Energy costs for pressurized fluidized bed combustion combined-cycle power plant systems are estimated to be up to 20% less than those for a conventional plant with stack gas cleaning at around $100/kW.

These savings are based on a 3% sulfur fuel cost of 80¢/GJ (10^6 Btu). An energy cost reduction of 17% is projected for a pressurized fluidized bed boiler (PFBB) operating at 17.5% excess air, with three stages of high-temperature particulate removal utilizing a calcium/sulfur ratio of 1.2 for a once-through sulfur removal system.

The same system with low-temperature particulate removal (e.g., water scrubbing) would be approximately 15% lower in energy costs than the conventional plant. This is a particularly significant result since it indicates that low-temperature particulate control, which may offer improved power plant reliability, is also economically attractive.

The pressurized fluidized bed boiler system, operating at 100% excess air to achieve higher carbon utilization in the primary combustors and greater turndown flexibility, is approximately 13% lower in cost but must utilize high-temperature particulate removal to be economical.

Energy costs for a pressurized fluidized bed adiabatic combustor combined cycle power plant with a once-through sulfur removal system using a calcium/sulfur ratio of 1.2 and three stages of particulate control are estimated to be approximately 7% lower than those of a conventional plant. Plant costs projected for regenerative fluidized bed combustion systems indicate regeneration is not cost competitive with the conventional plant until sorbent costs (including disposals) exceed $10 to $20/Mg.

A regenerative process is clearly attractive environmentally and, perhaps will be economically if a regenerative process is developed to operate with calcium/sulfur make-up ratios significantly less than 1.

REFERENCES

(1) D.L. Keairns, W.C. Yang, J.R. Hamm and D.H. Archer, "Fluidized Bed Combustion Utility Power Plants - Effect of Operating and Design Parameters on Performance and Economics," *Proceedings of the Third International Conference on Fluidized Bed Combustion,* Hueston Woods, Ohio, 1972.

(2) National Air Pollution Control Association - National Coal Board, Information Exchange Meeting in United Kingdom, April 20-24, 1970.

(3) H.R. Hoy and J.E. Stantan, *Amer. Chem. Soc. Div., Fuel Chem. Prepr.,* 14(2):59, 1970.

(4) *Pressurized Fluidized Bed Combustion,* Report No. 85, Interim No. 1, Office of Coal Research, National Research Development Corporation, London, England, 1974.

(5) *Reduction of Atmospheric Pollution,* Final Report, Vols. 1-3, Office of Air Programs, National Coal Board, London, England, September 1971, PB 210 673, PB 210 674 and PB 210 675.

(6) *Energy Conversion from Coal Utilizing CPU 400 Technology,* Interim Report No. 1, Office of Coal Research, Combustion Power Company, November 1974.

(7) R.G. Rincliffe, "The Eddystone Story," *Electrical World,* March 11, 1963.

(8) H. Harboe, "Coal for Peak Power," Stal-Laval, Ltd. (Presented at ACS National Meeting, Chicago, August 1973), *Div. of Fuel Chem. Preprints,* 18 (4).

(9) A.P. Fraas, *Potassium-Steam Binary Vapor Cycle with Fluidized-Bed Combustion,* (Presented at Annual AIChE Meeting, New York, November 1972).

(10) R.S. Rossbach, *Final Report of Joint NASA/OCR Study of Potassium Topping Cycles for Stationary Power,* GESP 741, NASA Lewis, General Electric Company, Cincinnati, Ohio, November 13, 1973.

(11) A.P. Fraas, *Comparison of Helium, Potassium, and Cesium Cycles,* Paper presented at the 10th Intersociety Energy Conversion Engineering Conference, Newark, Delaware, August 17-22, 1975.

(12) A.P. Fraas, R.S. Holcomb, M.E. Lackey and J.J. Tudor, *Design Study for a Coal Fueled Closed Cycle Gas Turbine System for MIUS Application,* Presented at the 10th Intersociety Energy Conversion Engineering Conference, August 17-22, 1975.

STUDIES OF ALTERNATIVE CONCEPTS AND MODIFICATIONS

The material in this chapter was excerpted from BNL 19308; PB 231 162; and PB 246 116. For a complete bibliography, see p 263.

FB ADIABATIC COMBUSTOR COMBINED CYCLE POWER PLANT

Adiabatic Combustor Concept

A pressurized fluidized bed boiler combined cycle power plant was designed using state-of-the-art power generation equipment. Performance, costs, and pollution abatement were projected for the system. The results show the concept has the potential to meet SO_2, NO_x, and particulate emission standards and may reduce energy costs 10% below a conventional plant with stack gas cleaning. This is achieved by effectively combining the combustion, heat transfer, and pollution control processes. The fluidized bed boiler operates at excess air values from 10 to 100%, with up to 70% of the heat released in burning the fuel with air transferred to the water/steam in the tubes surrounding and submerged in the bed.

Increasing the design point excess air with constant bed temperature will decrease the total heat transfer surface in the fluid bed until no boiler tube surface will be required at an excess air of approximately 300%. In this case, the power system is a combined cycle plant with the gas to the turbine expanders supplied from a coal-fired, adiabatic combustor (Figures 4.1 and 4.2).

Combined cycle plants of this type, which burn natural gas and/or heavy distillates, are now being marketed to electric utilities for intermediate and base load applications. One embodiment of such a plant is the Westinghouse Power at Combined Efficiencies (PACE) plant. In the PACE plant, the exhaust from the gas turbine is reheated to a temperature of about 1200°F ahead of the heat recovery boiler which has steam conditions of 1,250 psia/950°F.

A modified PACE configuration, shown in Figure 4.3, was selected for evaluation

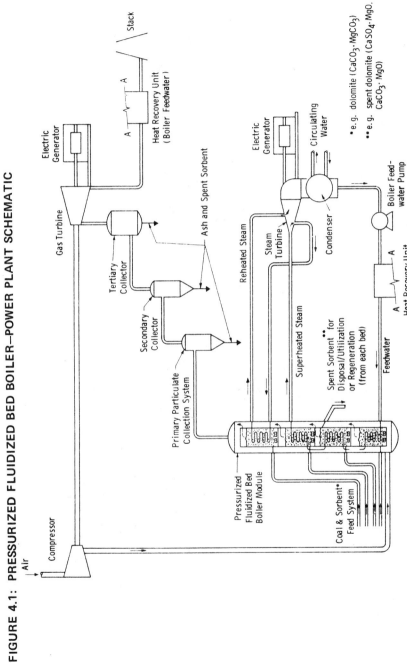

FIGURE 4.1: PRESSURIZED FLUIDIZED BED BOILER–POWER PLANT SCHEMATIC

Source: PB 246 116

FIGURE 4.2: PRESSURIZED FLUDIZED BED ADIABATIC COMBUSTOR—COMBINED-CYCLE POWER-PLANT SCHEMATIC

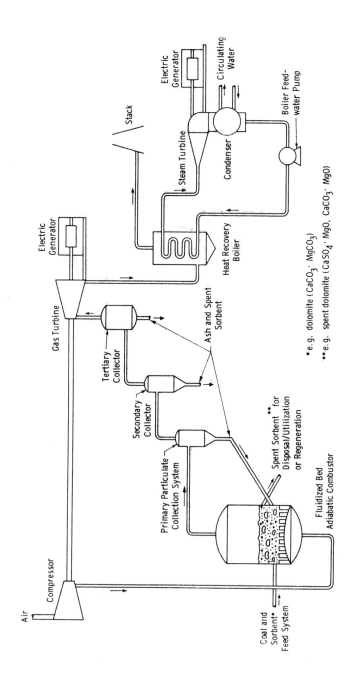

*e.g. dolomite ($CaCO_3 \cdot MgCO_3$)

**e.g. spent dolomite ($CaSO_4 \cdot MgO$, $CaCO_3 \cdot MgO$)

Source: PB 246 116

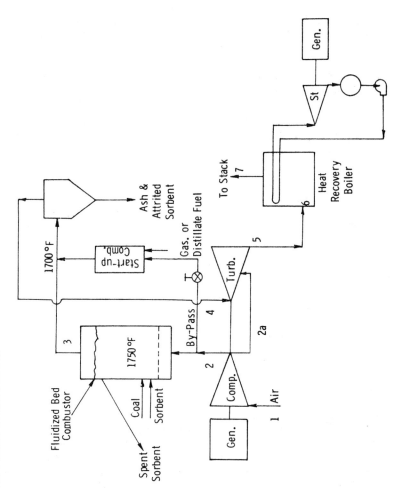

FIGURE 4.3:   COAL-FIRED PACE PLANT

of an adiabatic combustor system. In this configuration, the heat recovery boiler is unfired and desulfurization is achieved by feeding limestone or dolomite to the adiabatic combustor with once-through sorbent utilization.

Cycle Performance

The generating and performance data for an unfired heat recovery boiler case were selected for evaluation. As in the high-pressure fluidized bed boiler, the desulfurization process limits the maximum temperature in the fluidized bed to about 1750°F, which means that the maximum gas-turbine design temperature is about 1700°F. With a 1700°F gas-turbine inlet temperature, the temperature of the exhaust gas to the heat recovery boiler is about 850°F. This precludes the use of the standard PACE plant steam conditions.

The steam conditions selected for the adiabatic combustor system are approximately 700 psia/800°F. Also, since the gas temperature to the boiler will decrease with the gas-turbine load, it will be necessary to use a sliding pressure boiler in which the steam conditions will follow the exhaust gas temperature at part load. The estimated heat rate at the design point for this modified PACE plant is 9096 Btu (HHV)/kWh.

The standard PACE plant consists of two W501 gas turbines with a modularized heat recovery boiler for each gas turbine. With the gas turbine and steam conditions described above, this modified PACE plant would have net power output of 200.4 MW. In order to compare this concept with the high-pressure fluidized bed boiler with a capacity of 635 MW, a plant consisting of three coal-fired PACE systems was assumed with a total capacity of 601.2 MW.

A comparison of heat rates on the same basis for the various coal burning power plants is as follows:

Type of Plant	Heat Rate, Btu (HHV)/kWh
Conventional Steam	9186
High-Pressure Fluidized Bed Boiler	8990
Coal-Fired Modified PACE Plant	9096

Plant Turndown

The turndown capabilities of the coal-fired modified PACE plant should be very wide. A reduction of about 40% is possible by variation of bed temperature from 1750° to 1400°F (corresponds to turbine inlet temperatures of about 1700° and 1350°F). Modulation of compressor air provides another 20% turndown. Beyond this point, compressor air can by-pass the combustor (see Figure 4.3) operating at constant bed temperature to reduce the gas-turbine inlet temperature to the idle level of about 950°F.

It would probably not be practical to operate the boiler all the way to this point. Therefore, the boiler could be shut down and the plant power reduced to near zero. A by-pass around the boiler might be necessary. In order to carry out the by-pass technique described above, the design superficial velocity would have to be high enough to permit about 30% reduction in flow at a bed temperature of 1400°F.

Environmental Considerations

The environmental impact of an adiabatic fluidized bed combustor power plant is projected in Table 4.1. Sulfur dioxide emission standards can be met based on the available pressurized fluidized bed boiler data and the kinetic data at the adiabatic combustor operating conditions. Determinations must be made to ascertain if the nitrogen oxide emission standard can be met. Nitrogen oxide limits can be achieved in a pressurized fluid bed boiler. However, the high excess air for an adiabatic combustor may affect the NO formation. For example, if NO is reduced by reaction with CO, the high excess air may reduce the available CO and increase the NO emission.

Particulate emission standards can be achieved in an adiabatic combustor system based on fluid bed combustion emission and particulate removal equipment data. Data on particulate control for an adiabatic combustor system can confirm this projection. The gas flow for an adiabatic combustor system is ~40 lb gas/lb fuel compared with ~12 lb gas/lb fuel for a pressurized fluid bed boiler system. This requires a more extensive particulate removal system for the adiabatic combustor system than for a pressurized fluid bed boiler system.

Heat rejection with an adiabatic combustor system will be reduced ~35% below a conventional coal-fired plant. This results from the increased fraction of gas-turbine power. Solids from the system are projected from the pressurized boiler and TG data. Low-grade fuels and solid wastes can be effectively utilized in a fluidized bed adiabatic combustor system.

TABLE 4.1: ENVIRONMENTAL IMPACT

Item	Impact
Air Emissions	
SO_2, lb/10^6 Btu	< 1.2
NO_x, lb NO_2/10^6 Btu	to be determined
Particulate, lb/10^6 Btu	< 0.1
Heat rejection to cooling water	~35% less than conventional coal plant
Solids	dry, granular $CaSO_4$
Resources	multifuel capability, e.g., low-grade coals, oils, solid wastes

Source: PB 231 162

Fuel Processing Equipment

A preliminary evaluation was made of the adiabatic combustor fuel processing system. The design basis is as follows. There is 30 mol percent calcium utilization of the dolomite; 90% of the SO_2 is removed from burning Pittsburgh No. 8 coal of 4.3% sulfur; the average residence time of dolomite is 3 hours; the refractory insulation is 1700°F normal and 200°F design and the change in pressure across a wall is minimum (0.5 psi); the dolomite feed rate is 179,000 lb per hour (as $CaCO_3 \cdot MgCO_3$); and the coal feed rate is 140,000 lb per hour. Two design concepts—a single fluid bed module and a stacked fluid bed module—were considered as illustrated in Figure 4.4. A summary of the respective design

FIGURE 4.4: ADIABATIC COMBUSTOR DESIGNS

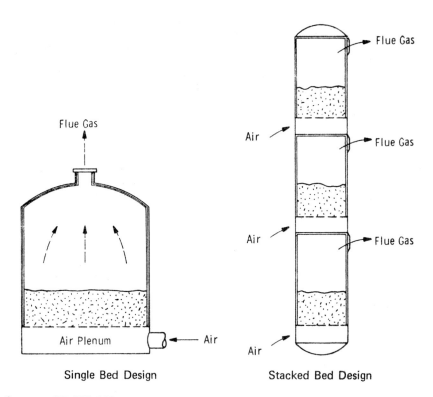

Single Bed Design   Stacked Bed Design

Source: PB 231 162

features is presented in Table 4.2. The particulate removal system design is specified in Table 4.3.

TABLE 4.2: ADIABATIC COMBUSTOR DESIGNS

	Single Bed Design		Stacked Bed Design	
	Design I	Design II	Design I	Design II
Number of Modules	4	2	4	2
Number of Beds per Module	1	1	3	6
Module Diameter, ft	21	30	12	12
Module Height, ft	16	16	50	100
Bed Depth, ft	6.5	6.5	6.6	6.6
Bed Area, ft^2	347	695	113	113
Fluidizing Velocity, ft/sec	6	6	6.2	6.2
Equipment Cost, $/kw	2.5	2.1	2.0	2.1

Source: PB 231 162

TABLE 4.3: PARTICULATE REMOVAL SYSTEM (200 MW)*

	1st Stage	2nd Stage
Actual gas flow, ACFM/module	124,500	124,500
Collector selection	Size 635, Model 810 Duclone	Model 18,000 Tornado Cyclone
Number of collectors per module	5	4
Pressure drop, inches w.g.	20.9	31
Collection efficiency, wt %	85.1	97

*This table is for the four-module design. For the two-module design, double the number of cyclones per module.

Source: PB 231 162

Power Plant Cost

The fluidized bed boiler power plant cost breakdown was reviewed and compared with an adiabatic combustor power plant. It was concluded that:

....Equipment costs for an adiabatic combustor power plant will be less that 4% below those of a conventional plant with stack gas cleaning using the same plant design basis.

....Total capital cost will be 5 to 15% less than that of a conventional plant with stack gas cleaning using the same plant design basis and construction time from 2½ to 3 years.

....The capital cost of the pressurized fluid bed boiler plant is up to 20% less than that of the adiabatic combustor plant.

....Plant heat rate can be reduced ~1% compared with conventional coal-fired steam plant using state-of-the-art technology.

....Environmental constraints can be met, heat release to cooling water is reduced ~35%.

....The simplicity of the adiabatic combustor system, the potential for a shorter construction time and the potential for near term commercial application may make the concept commercially viable.

RECIRCULATING BED BOILER DESIGN

A deep recirculating fluidized bed combustion boiler was conceived as an alternative to the proposed pressurized fluid bed combustion boiler concept. A recirculating bed boiler may offer advantages over the proposed design but requires more development and is thus considered a second generation concept.

The recirculating fluid bed concept is illustrated in Figure 4.5. Gas is fed to the base of an open draft tube section. The superficial velocity of the gas flowing up the riser may be 10 to 60 ft/sec. The solids are picked up pneumatically

FIGURE 4.5:   RECIRCULATION BED CONCEPT

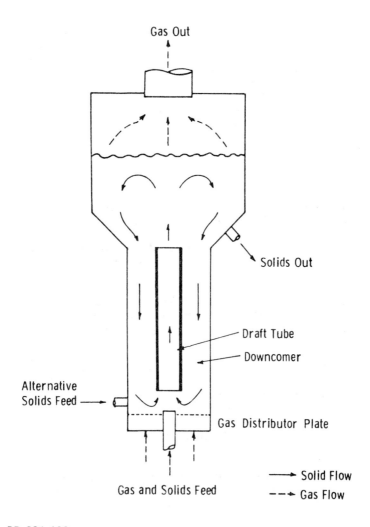

Source: PB 231 162

in the draft tube. The effective overall density of solids and gases is less in the draft tube section than in the downcomer, which creates a solid circulation pattern upward through the draft tube and downward in the downcomer. Solids and gases from the draft tube pass into a fluid bed which may be of expanded cross section. Solids from the fluid bed above the draft tube flow into the downcomers and enter the base of the draft tube. Gas is introduced at the base of the downcomer at a rate necessary to permit the downward flow of solids.

The potential advantages of a recirculating fluid bed boiler concept are:

....Feeding reactants to the bed. Oil and coal have been fed to recirculating beds without distribution and agglomeration problems.

....Boiler turndown and start-up. Locating heat transfer surface in the recirculating solids permits turndown by regulating the solids recirculation rate. Boiler start-up and shut-down are also simplified by so locating heat transfer surface.

....SO_2 removal and NO_x reduction. Limestone or dolomite can be used as the primary bed material in order to remove SO_2 and to catalyze the decomposition of NO. The riser portion of the bed can be operated under reducing conditions; the main bed and recirculating sections, under oxidizing conditions. In this way the formation of NO_x is reduced and the activation of the sorbent is promoted. It is postulated that fluctuation between oxidizing and reducing conditions promotes penetration of the limestone sorbent by the sulfur in the forms of sulfate, sulfite, or sulfide (1).

....Decreasing the cost of fluidized bed boilers further. A system study of pressurized fluidized bed boiler power plants indicated that they are lower in cost, more efficient in operation, and more effective in pollution abatement than conventional coal- or oil-fired power plants with stack gas scrubbing systems. The deep recirculating beds promise to reduce this cost further because of potentials in decreasing coal feed points, in increasing the heat transfer coefficient in the downcomer, and in simplifying the control and design of the boiler.

Several recirculating fluid bed systems have been developed and operated. The British Gas Council (2)(3)(4) developed deep recirculating fluidized beds for oil and coal gasification. In the gasifier this was used to smooth out the local high temperatures at the point of oxygen entry and in the hydrogenator to allow for the quick removal of particles wetted by the oil in the vicinity of the oil inlet and thus avoid the formation of agglomerates as well as closely control temperature for process purposes. Experiments were performed in large-scale models up to 5 feet in diameter and operated at pressures and temperatures up to 70 atm and 750°C (1382°F).

The same concept was also used for vertical pneumatic transportation of sticky and bridging powders (5). A variation of this concept was utilized to promote solid mixing, circulation, heat and mass transfer in the bed (6), and for pretreatment of caking coal (7). Taskaev and Kozhina (8) utilized a recirculating bed for the low-temperature carbonization of coals. Product gas preheated to a temperature of 600° to 700°C (1112° to 1292°F) was recirculated to entrain the coal particles of sizes up to ¼ inch through a vertical draft tube one inch in diameter and 20 inches in length into a concentric tube 6 inches in diameter and 40 inches in length. The coal or char particles fell through this annular space (a downcomer) in a dense descending bed and were then reentrained by the gas and recirculated until carbonization was complete.

A recirculating fluid bed reactor is utilized in the multistage fluidized bed coal gasification process for production of low Btu gas for power generation being developed by Westinghouse (9). A recirculating bed operating near 1600°F is used in this concept to devolatilize the coal and potentially to simultaneously desulfurize the fuel gas.

A preliminary design and evaluation of a pressurized recirculating fluid bed boiler for combined cycle power generation was made. The available literature on recirculating fluid beds was reviewed, cold model studies performed to establish initial operating and design criteria, a preliminary design developed, costs projected, and the concept evaluated. Therefore, a deep recirculating fluidized bed appears to be an attractive approach to the design of a boiler.

Deep Recirculating Fluidized Bed Boiler Concept

The concept of a deep recirculating fluidized bed boiler is illustrated in Figure 4.6. Primary air along with coal or oil fuel is fed to the boiler at the base of an open riser section. This section is bounded by metal or ceramic walls which may contain water-filled tubes for wall cooling and/or steam generation. The superficial velocity of the air and combustion gases flowing up the riser is 10 to 60 ft/sec at the operating temperature and pressure—1300° to 1950°F and 4 to 30 atm.

Coal fuel may either be pulverized (less than 200 mesh) or coarsely crushed (perhaps 1/4" x 0). Oil may be introduced to the boiler so as to wet the stream of sorbent after it flows down the downcomer, as it enters the base of the riser and jets upward in the riser. The fuel combines with the primary air in the riser. The heat produced increases the temperature of the combustion products, fuel ash, and sorbent; some heat may be removed through the walls of the riser. The quantity of sorbent entering the base of the riser is controlled by choice of design and operating parameters in order to limit the temperature increase of the gases and solids flowing up the riser.

The solids and gases emerging from the riser pass into a bed of expanded cross section and decreased superficial gas velocity. Baffles and/or heat transfer surface may be located in the bed section in order to obtain smooth fluidization and recirculation of the sorbent, to improve sulfur removal and NO reduction by the sorbent, and possibly to recover some heat. The gas velocity in the bed is in the range 1 to 15 ft/sec, sufficiently large to fluidize the sorbent but sufficiently small to minimize entrainment. The sorbent may either be coarsely crushed (1/4" x 0) or finely ground (perhaps 100 to 1,000 μ).

Hot sorbent from the bed flows into the downcomer, passes downward through the heat exchange surface (water/steam filled tubes which may, for example, be platens of horizontal or long vertical tubes), and enters the base of the riser section. The exchange surface extracts the heat energy transferred to the sorbent by the combustion process occurring in the riser and bed. Secondary air is introduced at the base of the riser at a rate necessary to permit the downward free flow of the sorbent and to prevent reducing conditions in the riser due to any residual fuel accompanying the sorbent.

Flow of the sorbent in a section of the downcomer can be reduced or halted by reducing or cutting off the flow of secondary air to that section. This capability makes it possible to adjust independently heat removal and heat production in the boiler. Control of the primary air can be used to adjust the recirculation

FIGURE 4.6:   DEEP RECIRCULATING FLUIDIZED BED BOILER

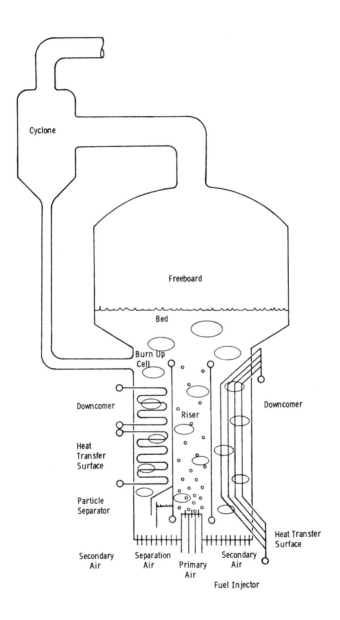

Source: PB 231 162

rate of the solids and the degree of oxidation in the riser and above the bed. Combustion after the bed increases the range of gas-turbine operation. Start-up and turn-down are thus facilitated.

Cold Model Studies

In order to study the feasibility of using deep recirculating fluidized beds for coal combustion boiler power plants, a two-dimensional cold model was constructed from the transparent acrylic sheets to investigate solid circulation, operation characteristics, and design criteria. Only the bed cross section, 8.5" x 1.5", was fixed; the other dimensions were adjusted to study the effect of these design variables. Pressure taps were also provided for measuring differential pressure drops.

Some experiments were performed in a slightly modified two-dimensional bed. Two separate fluidized beds (or "boots") were created by inserting two aluminum bars ¾" thick underneath the draft tube. The objective of this modification was to study the effect of the draft tube inlet design on the solid circulation rate and the air distribution between the downcomers and the draft tube.

Serpentine solid copper rods were also inserted in the downcomer sections to simulate the imbedded heat transfer surfaces. The serpentine rods are supported by inserting the ends of the rods into closely sized holes on the aluminum bars. Two serpentine rods were arranged in parallel 4.5 inches from the top of the draft tube with equal distances between the rods and between the rods and the walls.

Ottawa sand was used as the bed material. During operation, air is injected into the draft tube through the air jet nozzle to provide an air velocity up to ~43 ft/sec in the draft tube while the downcomer sections are "minimally fluidized." This creates a solid circulation pattern upward through the draft tube and downward in the downcomers. The minimum fluidizing condition in the downcomers is assured first by turning up the air supply to the downcomers until air bubbles appear in the downcomers, and then turning down the air supply until air bubbles just disappear.

The solid downward velocity in the downcomer was estimated by following a tracer particle for 16 inches with a stopwatch. The tracer particles are silica gel dyed with red pigment and of similar size. For every operating condition, at least ten particles in each downcomer were traced and arithmetic average velocity was taken to be the solid particle velocity in the downcomer. The solid recirculation rate was then calculated by assuming plug flow in the downcomer.

A mathematical model has been developed to predict solid circulation rate (10). The model is good to within ± 20% if experimental values of slugging pressure losses are used. The critical design variables for a recirculating fluidized bed boiler are identified as the draft tube inlet design, the draft tube height, and the downcomer/draft tube area ratio.

Bed-Tube Heat Transfer Coefficient

No heat transfer experiment was performed to obtain the bed-tube heat transfer coefficient for the boiler tubes immersed in the downcomer; however, literature data on heat transfer coefficients in a moving bed were utilized to project a reasonable value for this recirculating bed boiler design. Heat transfer in a gas fluidized bed was reviewed; heat transfer data in a moving bed has been utilized.

If radiative heat transfer can be ignored, the dominating factor in the heat transfer is the high heat capacity of solid particles relative to that of gas. Heat is primarily transferred from the bulk of the bed by solid convection, and convective heat transport through the gas is relatively unimportant. Because of the high surface area exposed by the particles, the solids effectively act as a local heat source or heat sink. Consequently, the thermal gradient between the immersed transfer surface and the bed is restricted to the zone immediately adjacent to the surface.

Thus highest overall heat transfer coefficient will occur under conditions which limit particle residence times at the surface. It is difficult to rely on the ordinary bubbling processes in a fluidized bed for controlling the heat transfer in the bed. A forced particle circulation such as in a recirculating bed design will be able to control the particle flow past the heat transfer surfaces in a regulated manner.

Botterill et al (11)(12)(13)(14) examined the relation of heat transfer efficiency with particle residence time on the heat transfer surfaces, both experimentally and theoretically, and found that high rates of heat transfer could be obtained if particle residence time at the transfer surface was sufficiently limited. The experiments were carried out in a stirred packed bed (11)(12), and a flowing bed with a small exposed wall heat transfer surface. The bed-tube heat transfer coefficients obtained ranged from 70 to 280 Btu/ft^2/hr/°F depending on particle sizes, particle properties, and particle residence times at the heat transfer surface. Actual heat transfer coefficients for various densities of fluidized coal were also measured. The results ranged from 80 Btu/ft^2/hr/°F for a residence time of ~50 milliseconds to 190 Btu/ft^2/hr/°F for a residence time of ~7 milliseconds.

Experiments were also carried out by mounting arrays of tubes on suitable jigs in the straight section of the continuous channel containing a flowing fluidized bed (14)(15). Both horizontal and vertical tube bundles were used in the experiments. Local heat transfer coefficients between elements of the surface and the bed in cross flow increased with an increasing solids flow rate, as expected. The local heat transfer coefficients were between 100 and 130 Btu/ft^2/hr/°F. However, when the solids flow rate was increased, a wake developed from the downstream-facing surfaces, and there was a marked tendency for the bed to defluidize locally at the upstream-facing surface. Consequently, the overall bed-tube heat transfer coefficient was not very sensitive to solids flow rate and was of the order of 55 Btu/ft^2/hr/°F. There was no attempt to align the axes of the tube bundles in the flow direction of the flowing fluidized bed.

Another similar heat transfer phenomenon is that of heat transfer between submerged heat transfer surfaces and the bed in a spouted bed. The heat transfer coefficient measured in the annular section between the spout and the containing wall was found to change from ~30 Btu/ft^2/hr/°F close to the wall to ~50 Btu/ft^2/hr/°F close to the spout (16). The only difference from the heat transfer in the downcomer section of a recirculating bed is that the downcomer section is minimally fluidized in the former design while there is no gas input to the annular section in the latter case. Due to this added agitation, the heat transfer coefficient in a recirculating bed is probably higher than that in a spouted bed.

Zenz and Othmer (17) reviewed the heat transfer in a moving bed. The typical heat transfer coefficient obtained for coal and limestone in a horizontally flowing bed of solids ranges from 40 to 100 $Btu/ft^2/hr/°F$. For a moving bed with little relative movement between particles, the thermal conductivity of the solid particles becomes the controlling factor. Equations are available for these calculations.

Based on this literature review, a commercial recirculating bed boiler was designed by assuming the overall heat transfer coefficient in the downcomer section was 50 and 75 $Btu/ft^2/hr/°F$. The boiler costs are then compared with that of the basic design.

Conceptual Recirculating Bed Boiler Design

A conceptual recirculating bed boiler design was prepared on the basis of the boiler plant design developed by Westinghouse under contract to EPA (9). The operating conditions and design parameters for a 318 MW boiler and the power cycle were summarized and the conceptual recirculating bed boiler design evaluated in two cases.

Case 1 assumes the overall heat transfer coefficient in the downcomer section to be 75 $Btu/ft^2/hr/°F$; Case 2 assumes 50 $Btu/ft^2/hr/°F$. The coal is assumed to combust mostly in the draft tube and partially in the fluid bed above the draft tube. The heat transfer surfaces in the fluid bed keep the bed temperature essentially at 1750°F. The heat transfer surfaces in the downcomer section reduce the solids temperature from 1750°F at the top of the downcomer down to ~1000°F at the bottom of the downcomer. The solids are then picked up in the draft tube to increase their temperature to 1750°F.

The design selected for a 318 MW recirculating bed boiler consists of six individual modules of 12 ft inside diameter with one module for preheater, two modules for evaporator, two modules for superheater, and one module for reheater. This design was selected to permit standard shop fabrication of each vessel. A four-module design could also be developed, which would increase the diameter of two of the reactors. The conceptual design is depicted in Figure 4.7.

Coal is fed concentrically with 90 to 95% stoichiometric air into the draft tube. Most of the coal is expected to be consumed in the draft tube. The remaining coal is carried into the expanded fluid bed above the draft tube where additional air is available from the downcomer section. Thus the draft tube is operated under a slightly reducing condition while the fluid bed above the draft tube is operated under an oxidizing condition with 10% excess air.

High excess air can be utilized in this design. However, high excess air (e.g., 100%) has not been considered in the present evaluation. The effect on vessel design and performance (e.g., turndown, NO_x) would have to be considered. The bed cross section expands at the outlet of the draft tube to reduce the superficial gas velocity to 3 to 4 ft/sec to prevent slugging and to reduce carryover of solid particulates. The heat transfer surfaces are vertical tubes of 2 inches

FIGURE 4.7: SCHEMATIC DESIGN OF THE ADVANCED RECIRCULATING
BED COMBUSTOR

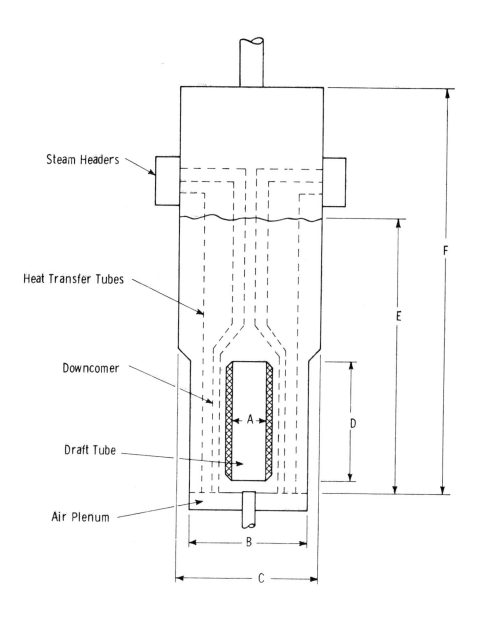

Source: PB 231 162

in diameter and extend through the downcomer section and the fluid bed above it. The horizontal bends into the steam headers just above the fluid bed will help prevent slugging and reduce elutriation. The largest vessel diameter was restricted to 12 ft inside diameter so that the vessels can be shop fabricated and shipped by rail.

Boiler Operation and Performance

The coal-burning recirculating fluidized bed boiler would be operated somewhat differently from the basic pressurized fluidized bed boiler because of the differences in boiler design and modular arrangement. In the recirculating fluidized bed boiler design, each individual module performs only one function, i.e., it is either for preheating, evaporating, superheating, or reheating. Thus, this concept requires a minimum of four modules to complete the total boiler function from preheating to superheating and reheating.

In the pressurized fluidized bed boiler design, however, each module contains one preheater bed, two superheater beds, one reheater bed, and walls for evaporating. Thus, each module is capable of generating 280 MW electrical power. Due to this difference in modular arrangement, the philosophy for cold start-up, hot restart, turndown, and load control would be different.

The results from the cold model studies in the two-dimensional model indicate that the solid circulation rate in a recirculating bed can be easily controlled by turning down the air input into the downcomer section. The circulation can be effectively stopped by completely shutting off the air input into the downcomer. Thus, theoretically, the recirculating fluidized bed boiler will be able to follow the load change continuously from no load to full load.

Subdivision into modules for turndown consideration is not necessary. The design of the recirculating bed boiler is thus based on other considerations, such as ease in assemblage and erection, control and turndown, and economics. To be conservative, the 318 MW recirculating fluidized bed boiler design consists of six modules: one for preheating, two for evaporating, two for superheating, and one for reheating. The downcomer section of preheater and reheater modules is separated into two sections by water walls to facilitate turndown. Air plenums for evaporator and superheater are sectioned into several sections so that the circulation rate of each section can be individually adjusted.

The bed temperature in the fluid bed above the draft tube is the primary variable for control: too high a temperature will cause ash to fuse and dolomite to lose reactivity; too low a temperature will decrease substantially the reactivity of dolomite and possibly decrease the combustion efficiency of coal in the bed. If carbon is elutriated from the bed, it can be recycled and fed at a point just above the downcomer section so that the elutriated carbon particles can be trapped in the descending bed in the downcomer to increase residence time of the particles.

There is no fresh coal feed in the downcomer section to compete for the air, and there is an additional chance for the unburned carbon particles to be picked up in the draft tube and carried into the fluid bed above the draft tube. Thus, the carbon burn-up cell in the pressurized fluidized bed design (a carbon burn-up unit is not considered necessary in either design with high excess air) probably is not needed in the recirculating fluidized bed design, which would simplify the control scheme and boiler design.

Air input into the downcomer section will be used to control the solid recirculation rate and thus the heat transfer rate to the heat transfer surfaces. Because of this great flexibility, cold start-up, hot restart, and load control are much simpler. To start up, auxiliary fuel can be fed into the fluid bed through the draft tube. When the bed temperature reaches the coal ignition point ($\sim$1000°F), coal feed can be started. The solid circulation can be started when the bed temperature reaches the design temperature (1750°F). For hot restart, the recirculating bed design would be even more convenient, because the bed is not completely shut off during turndown as in the case of the pressurized fluidized bed design.

The bed is essentially maintained at the operating temperature at least for some sections. A hot restart will simply mean an increase in the coal feed rate and the solid circulation rate, while in the pressurized fluidized bed design, the bed has to be brought up to the coal ignition temperature by auxiliary firing before coal feeding can be initiated. This means an additional time delay in following the load change. This potential capability of shortening the load response time is an important advantage for the recirculating fluidized bed design.

The performance characteristics of the recirculating fluidized bed should not be too different from those of the basic pressurized fluidized bed design. The overall boiler efficiency is expected to be comparable, with slightly higher pressure drop through the recirculating fluidized bed due to deeper bed design; however, the air pollution control capability in the recirculating bed is potentially greater. The two-stage combustion of the coal in the recirculating bed—reducing conditions in the draft tube section and oxidizing conditions in the downcomer section and the main bed—should reduce the formation of NO_x. If fluctuation of the sorbent between oxidizing and reducing conditions promotes penetration of the sulfur into the sorbent in the form of sulfate, sulfite, or sulfide, then increased activity and higher utilization of the sorbent can be achieved.

Economics

In addition to the air pollution control capability, the recirculating fluidized bed boiler possesses the following potential advantages:

....in decreasing coal feed points. Coal can be fed and combusted in the draft tube through a single coal feed point. The dilute pneumatic transport of coal particles in the draft tube increases gas/solid contact. The air by-passing through bubbles which adversely affect the gas/solid contact in a fluidized bed does not occur in the draft tube.

....in preventing agglomeration. High solid circulation rate in the recirculating bed prevents caking and agglomeration of coal in the vicinity of coal feeding points.

....in capability of designing for deeper bed. The increased circulation and mixing in a recirculating bed eliminates the problems of

particle segregation and temperature gradient in the main bed. This
permits design for a deeper bed.

....in control and load response. Continuous turndown is possible in
a recirculating bed. For cold start-up or hot restart, a recirculating
bed is easier to handle. The response time is shorter in following
load change in a recirculating bed.

....possibly in increasing the heat transfer coefficient in the downcomer
section over that in a fluidized bed. Because of fast solid circulation
through the downcomer section, there is a possibility of increasing
the heat transfer coefficient in the downcomer section to higher than
that in a fluidized bed (50 Btu/ft^2/hr/°F), which must be verified experimentally.

All the advantages mentioned here can be directly or indirectly transferred into
economic savings in designing a recirculating fluidized bed boiler.

The design for the 318 MW recirculating fluidized bed boiler is a six-module plant
with one module for preheating, two for evaporating, two for superheating, and
one for reheating. Each module consists of only one deep recirculating bed as
compared to four stacked beds in the pressurized fluidized bed boiler design.

The pressurized vessels are designed for operation at maximum operating conditions of 150 psig and 650°F with 6-inch refractory insulation (4-inch insulation plus 2-inch hard face). Draft tubes are constructed from either 316 stainless steel or Incoloy 800 and are to be water- or steam-cooled. Vertical tubes
of conventional boiler tube materials are used for heat transfer surfaces. The
total cross section of the reactor vessel is used for the recirculating bed area as
compared to <40% utilization of vessel cross-section area in the pressurized
fluid bed design. The cost of the six-module boiler is evaluated for the heat
transfer coefficient in the downcomer section of 75 and 50 Btu/ft^2/hr/°F. The
heat transfer coefficient in the fluid bed above the draft tube is 50 Btu/ft^2/hr/°F.

Although the heat transfer surface requirements for the recirculating bed design
are 6 to 38% higher than those for the basic pressurized fluidized bed design,
the overall boiler costs may be lower by 5 to 13%. The increase in heat transfer surface requirements is due to a decrease in temperature difference between
the descending bed in the downcomer section and the heat transfer surface; however, the decrease in module height from four modules of 110 ft to six modules
of 30 ft substantially reduces the boiler costs.

This reduction in module height is primarily due to elimination of a stacked
bed design. When fluidized beds are stacked one on top of the other as in the
basic design, freeboard and air plenum have to be provided for each bed, thus
increasing the module height.

The cost of the particulate removal system should be comparable in both designs
because the airflow rate is similar and the dust loading should be similar as well.
Lower velocities in the freeboard area in the recirculating bed design should
result in slightly lower dust loading if slugging does not set in.

The structural cost for six modules, each 30 feet high, should be lower than that for four modules, each 110 feet high. The manifolding, however, would be more expensive for the 6-module design. The instrumentation and control are projected to be easier and less costly for the recirculating bed design. Only 6 beds are involved, as compared to 16 individual beds in the pressurized fluidized bed design. It requires turning on and off a complete fluidized bed or a complete module in the pressurized fluidized bed for turndown purposes, while in the recirculating bed design it only requires adjustment in downcomer airflow rate.

The coal feeding points are reduced from 64 to 6, which also substantially reduces the boiler costs. Thus, it is probably conservative to assume that the cost of auxiliary systems are comparable for both the recirculating bed and the pressurized fluidized bed designs.

Assessment

The overall boiler costs of a recirculating bed design are estimated to be lower by 5 to 13% than those of the pressurized fluidized bed basic design, depending on the construction of the draft tube and on the heat transfer coefficient in the downcomer section. The costs are arrived at by assuming that the costs of the auxiliary systems are comparable. This is a conservative assumption, in view of the simplicity in operating a recirculating bed boiler. In addition to the cost advantage, the recirculating bed design has the following potential:

....in preventing agglomeration. High solid circulation rate in the recirculating bed prevents caking and agglomeration of coal in the vicinity of coal feeding points so that low-grade coal can be used.

....in capability of deep bed operation. The increased circulation and mixing eliminates the problems of particle segregation and temperature gradient in the main bed. This permits design for a deeper bed.

....in control and load response. Continuous turndown without shutting off the bed is possible. This promises easier start-up and faster response in following load changes.

....in SO_2 removal and NO_x reduction. The two-stage combustion—operating the riser portion of the bed under the reducing conditions and the main bed and downcomer section under oxidizing conditions—reduces the formation of NO_x and promotes activation of the sorbent.

The recirculating boiler concept offers many potential advantages over the base pressurized fluid bed boiler design. Thus, a recirculating fluid bed boiler offers economical and environmental advantages as a second generation pressurized fluid bed boiler.

ROTATING FLUIDIZED BED REACTOR

Advantages for Power Generation

A fluidized bed combustion system, operating at elevated pressure in a combined cycle power plant, offers the greatest potential for producing electrical energy from fossil fuel within environmental constraints and at a cost less than conventional power plants using a low sulfur fuel or stack gas cleaning.

A pressurized fluidized bed combustor combined cycle can be operated with different configurations. The fluidized bed could be operated close to the stoichiometric requirement of air with the heat released in burning the fuel transferred to water/steam in the tubes surrounding and submerged in the bed.

Increasing the design point excess air with constant bed temperature decreases the total heat transfer surface in the fluid bed until no boiler tube surface will be required at an excess air of about 300%. In this case, the power system is a combined cycle plant with the combustion product gases as feed to the gas turbine. The exhaust from the gas turbine is then fed to the heat recovery boiler to run a steam turbine.

A rotating fluidized bed reactor could advantageously be utilized as a coal combustor for power generation. The device, based on technology developed at Brookhaven National Laboratory, utilizes a fluidized bed of coal which is made dense by centrifugal force acting on the particles as they rotate in a basket through which the fluidizing air passes. The rotating bed combustor produces a much higher volumetric energy release than is attainable in conventional fluidized bed systems and the inclusion of limestone and/or dolomite removes sulfur dioxide formed during combustion.

The principle of a rotating fluidized bed is shown in Figure 4.8. The principal feature is the rotating member which can be supported on appropriate bearings and rotated by mechanical means.

FIGURE 4.8: ROTATING FLUIDIZED BED

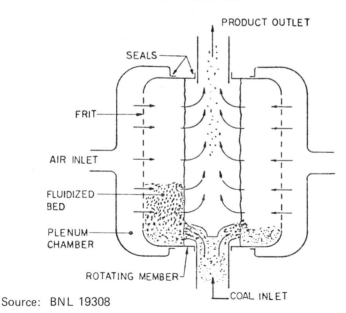

Source: BNL 19308

As a result of the rotation, a radial centrifugal force is developed which in a moderate-sized vessel of 5 ft i.d. with a bed depth of 1 ft reaches $40\,g$ at a rotational speed of 200 rpm. The centrifugal force essentially overrides the $1\,g$ earth gravity downward force and the bed material lines up against the outside wall of the vessel. The magnitude of the centrifugal force is also influenced by the particle mass and size of the bed material.

Air enters the plenum chamber as shown, and flows radially inward through an appropriate porous wall (called the frit) and reacts with and fluidizes the bed of finely-divided solids. The centrifugal outward force of the bed is essentially balanced by the ingoing pressure force of the air passing through the bed, so that the bed is supported by the air flow. The combustion product gas leaves the surface of the bed and curves towards the product outlet nozzle. Ash and unreacted carbon will also leave through the outlet nozzle due to the decrease in size or density of the particle. The spent stone will be separated from the gas at the bed exit.

Because of a high gravitation field created by rotation of the fluidized bed, much higher velocities can be used compared with a conventional fluidized bed. This results in higher heat and mass transfer coefficients which yield significantly improved performance and capacity characteristics compared with a $1\,g$ bed.

Process Design for Power Generation

A preliminary design was based on operating conditions and design parameters selected by evaluating available data on fluidized bed combustion and desulfurization, power cycles, and alternative boiler concept. Figure 4.9 presents the flow diagram of a combined cycle power generation system using a pressurized fluidized bed reactor.

Coal and dolomite are added to the rotating fluidized bed continuously. The stone is separated from the gas at the bed exit as a result of the centrifugal forces imparted by the rotation of the bed and the reduction in the gas velocity in the bed outlet duct. Ash is carried away with the product gas due to its small diameter and is separated in a cyclone separator. Product gas after cyclone separation is fed to the gas turbine.

The desulfurization process limits the maximum temperature in the fluidized bed to about 1750°F, which means the maximum gas turbine design temperature is about 1700°F. With a 1700°F gas turbine inlet temperature, exhaust gas from the turbine will be about 850°F and will be fed to a secondary heat recovery boiler, where steam will be generated at approximately 700 psi and 800°F.

The fuel selected for the rotary fluidized bed design is Ohio Pittsburgh No. 8 seam coal. The average coal particle size selected for design purposes is 500 microns. The sorbent used for desulfurization is BCR 1337 dolomite. The coal contains 3.3% by weight moisture, 71.2% carbon, 5.0% hydrogen, 6.4% oxygen, 1.3% nitrogen, 4.3% sulfur and 8.5 % ash. Its net heating value is 12,598 Btu per pound. Analysis of the dolomite is as follows (by weight percent): SiO_2, 0.78; Al_2O_3, 0.15; Fe_2O_3, 0.25; MgO, 45.00; CaO, 53.62; TiO_2, 0.02; SrO, 0.03; Na_2O, 0.02; K_2O, 0.10; MnO_2, 0.03. Its specific gravity is 3.476.

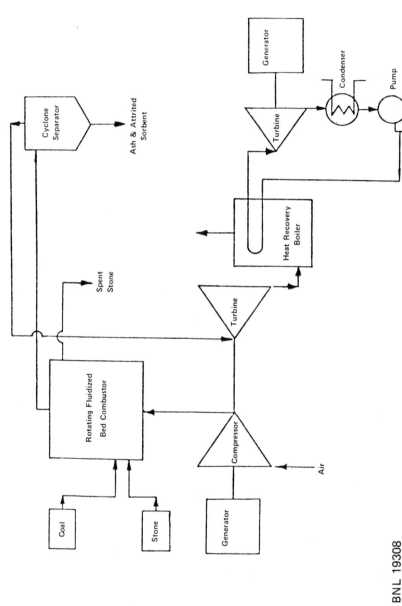

FIGURE 4.9: PROCESS DESIGN FOR POWER GENERATION

Source: BNL 19308

FIGURE 4.10: SYSTEM WITH WATER USED AS HEAT TRANSFER MEDIUM

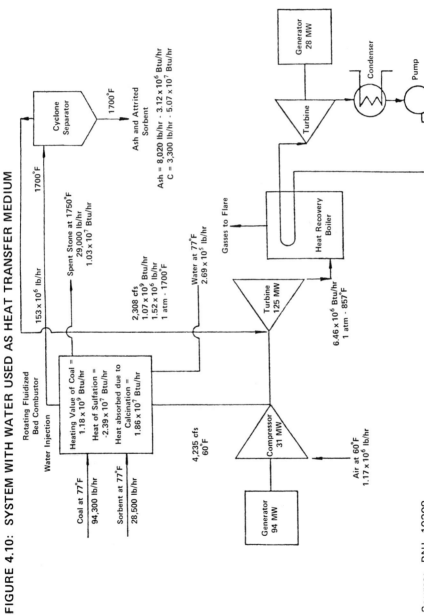

Source: BNL 19308

Uncalcined dolomite is fed continuously to the combustor. The average residence time of stone in the bed is one hour, and 1.5 times the stoichiometric requirement of stone is fed continuously to the combustor. The average stone size selected is 500 microns. The selection of 500-micron stone size and a Ca/S molar ratio of 1.5 is based on the assumption that 90% of the sulfur is removed. Since the rate of reaction increases with decrease in particle size, an even smaller particle size could be used by increasing the rotational speed of the bed.

The operating fluidization velocity based on the average particle size (500 microns) of the bed is selected sufficiently below the carryover velocity to prevent loss of the bed and sufficiently above the minimum fluidization velocity to insure fluidization of the bed.

Water Used as Heat Transfer Medium

In a pressurized system, the use of excess air to keep the bed temperature below 1750°F requires a significant amount of power to compress the inlet combustion air, thus markedly decreasing the net power generation and overall thermal efficiency.

Direct cooling with water using the latent heat of vaporization to decrease the flame temperature could be utilized in a pressurized fluidized bed combustor. No excess air would be required. Figure 4.10 shows the flow diagram of this system along with material and energy balances.

Water is fed to the rotating fluid bed along with air. Heat released in burning the fuel is transferred to the water. The dry, superheated steam produced is fed to the gas turbine along with combustion product gas. Exhaust gas from the turbine is fed to the secondary heat recovery boiler to generate power in the steam turbine. Reactions leading to the formation of H_2 and CO below 1750°F in this system, especially in the presence of excess air, are practically nonexistent (18). The operating conditions and performance of the system are summarized in Table 4.4.

TABLE 4.4: ROTATING FLUIDIZED BED COMBUSTOR PROCESS CONDITIONS AND PERFORMANCE*

Inside diameter of bed	6 ft
Length of bed	6 ft
Bed thickness	1 ft
Bed temperature	1750°F
Coal rate	94,320 lb/hr
Coal inlet temperature	77°F
Coal particle size	500 microns
Combustion efficiency	95%
Dolomite rate	29,000 lb/hr
Dolomite inlet temperature	77°F
Dolomite particle size	500 microns
Sulfur removal efficiency	90%
Ca/S molar ratio	1.5
Average residence time of dolomite	1 hr
Air rate	1.165×10^6 lb/hr

(continued)

TABLE 4.4: (continued)

Air inlet temperature to combustor	$60°F$
Air inlet pressure to combustor	10 atm
Compression ratio	10
Percentage excess air	30%
Compressor stages	3
Power to compressor	31 MW
Water rate	2.69×10^5 lb/hr
Temperature of gas to turbine	$1700°F$
Product gas rate	1.527×10^6 lb/hr
Exhaust gas from gas turbine	$857°F$
Net power generated by gas turbine	94 MW
Temperature out of recovery boiler	$408°F$
Power generated by steam turbine	28 MW
Total plant power	122 MW
Average velocity through the bed	7.25 ft/sec
Force of acceleration	$30\,g$
Rotational speed	170 rpm

*Desulfurization with dolomite and internal injection of cooling water, 10 atm.

Source: BNL 19308

REFERENCES

(1) ANL/ES/CEN-F026, *Monthly Progress Report #26,* December 1970, Argonne National Laboratory, Chemical Engineering Division.

(2) Horsler, A.G., Lacey, J.A., and Thompson, B.H. *Chem. Eng. Prog.,* 65 (10), 59 (1969).

(3) Dent, F.J. *Methane from Coal.* 9th Coal Science Lecture, BCURA (1960).

(4) Horsler, A.G., and B.H. Thompson. *Fluidization in the Development of Gas Making Processes.* Tripartite Chemical Engineering Conference, Montreal, Canada (1968).

(5) Decamps, F., Dumont, G., and Goossens, W. "Vertical Pneumatic Conveyer with a Fluidized Bed as Mixing Zone." *Powder Technol.,* 5, 299 (1971/72).

(6) Buchanan, R.H. and Wilson, B., "The Fluid-Life Solids Recirculator," *Mech. & Chem. Eng. Trans. (Australia),* 117, May (1965).

(7) Curran, G.P., and Gorin, E. *Studies on Mechanics of Flow Solids Systems.* Report prepared for Office of Coal Research by Consolidation Coal Company (1968).

(8) Taskaev, N.D. and Kozhina, M.I. *Trudy Akad. Nank Kirg. 2 S.S.R.,* 7, 109 (1956).

(9) Archer, D.H., Vidt, E.J., Keairns, D.L., Morris, J.P. and Chen, J.L.P. *Coal Gasification for Clean Power Production.* Proceedings of the Third International Conference on Fluidized Bed Combustion, Hueston Woods, Ohio, October 1972.

(10) Yang, W.C. and Keairns, D.L. *Recirculating Fluidized Bed Reactor Data Utilizing a Two-Dimensional Cold Model.* Paper presented at the 75th National Meeting, AIChE, Detroit, June 3-6, 1973.

(11) Botterill, J.S.M., Butt, M.H.D., Cain, G.L. and Redish, K.A. *Proceedings of the International Symposium on Fluidization.* Netherlands University Press, Amsterdam, 1967.

(12) Botterill, J.S.M., Butt, M.H.D., Cain, G.L., Chandrasekhar, R. and Williams, J.R. *Proceedings of the International Symposium on Fluidization,* Netherlands University Press, Amsterdam, 1967.

(13) Botterill, J.S.M., Chandrasekhar, R. and Van der Kolk, M. *Brit. Chem. Eng.* 15 (6), 769 (1970).

(14) Botterill, J.S.M. *Powder Tech.* 4, 19 (1970/71).

(15) Botterill, J.S.M., Chandrasekhar, R. and Van der Kolk, M. *Chem. Eng. Prog. Symp. Series* 66, No. 101 (1970).

(16) Mathur, K.B., Chap. 17, *Fluidization,* edited by J.F. Davidson and D. Harrison, Academic Press (1971).

(17) Zenz, F.A. and Othmer, D.F. *Fluidization and Fluid-Particle Systems.* Reinhold Publishing Corporation, N.Y. (1960).

(18) Haslam and Russell, *Fuels and Their Combustion,* McGraw Hill, New York, 1926.

SORBENTS
FOR SULFUR REMOVAL SYSTEMS

The material in this chapter was excerpted from PB 246 116. For a complete bibliography see p 263.

The sulfur removal system in fluidized bed combustion is based on the principle that a solid sorbent can trap the fuel sulfur in solid form as the coal is burned and prevent its release to the environment as gaseous sulfur dioxide (SO_2). Thermodynamic analysis shows which solids will react with sulfur dioxide under process conditions and, therefore, defines those sorbents which must be considered for use.

The calcium-based sulfur removal process has been developed more extensively than have those using alternative sorbents. Experimental work thus far has used limestone or dolomite sorbents as sources of calcium carbonate ($CaCO_3$) or calcium oxide (CaO) in the sulfur removal processes (Figure 5.1). Primary consideration has been given to optimizing operation of the fuel processing-sulfur removal system module.

More than 90% of sulfur dioxide emissions can be prevented and the fuel sulfur captured in a dry solid using limestone or dolomite as sorbents. The environmental standard of 0.54 kg (1.2 lb) sulfur dioxide/1.055 GJ (10^6 Btu) can be readily obtained. Maintenance of this standard of sulfur dioxide pollution abatement for the combustor operating conditions, while minimizing solid waste accumulation, has required excess sorbent or calcium/sulfur molar feed ratios of about 2/1 for dolomite and 3/1 for limestone. The reduction of these ratios is a major goal in the development of fluidized bed combustion systems.

SORBENT SELECTION

The establishment of stone selection criteria for choosing limestones and dolomites suitable for use as fluidized bed desulfurizing agents is important for optimizing the fluidized bed combustion process. Stone selection criteria should not be rigid specifications, but depend on the particular system design.

FIGURE 5.1: TEMPERATURE AND PRESSURE CONDITIONS FOR STABILITY OF THE SULFUR SORBENT AT PROJECTED COMBUSTOR OUTLET AIR COMPOSITIONS

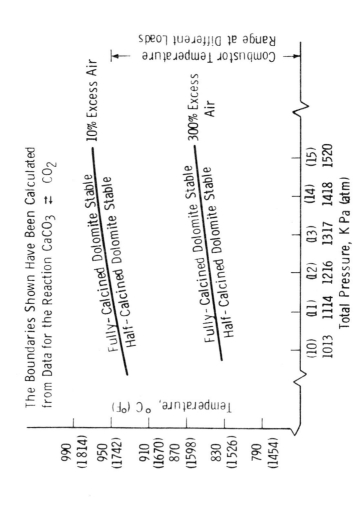

The use of calcium-based sorbents (limestone or dolomite) to trap in solid form the sulfur released from "dirty" fuels during combustion is based on the thermodynamic stability of calcium sulfate under fluidized bed combustion conditions; and on the kinetic efficiency, structural integrity, and economical availability of the sorbent, which permit application of the basic idea in a technically sound process.

Calcium carbonate is found as both limestone and dolomite in the eastern and midwestern states which produce high-sulfur coal. Both pure and impure limestones and dolomites have been tested as sorbents: the sulfur removal efficiency has generally been well within EPA limits. In order to select the best available material as sorbent, sorbent selection criteria must be developed. Establishing selection criteria minimizes the cost and time involved in assessing the usefulness of rock quarried near a particular plant site. The criteria for choosing a stone are based on:

Acceptor properties for the stone for sulfur removal
Attrition resistance of the stone
Trace element emission characteristics
Regeneration characteristics
Suitability of spent sorbent for final processing for disposal
Economic availability of the stone

Acceptor Properties of the Stone for Sulfur Removal

The acceptor properties of the stone depend on the stable form of the sorbent in the particular process, and on the kinetics of the sulfation reaction. Both limestone and dolomite may be used as sorbents at atmospheric pressure. At higher pressures and under exit gas conditions as Figure 5.1 shows (1) calcium carbonate is the stable form of the sorbent. Calcium carbonate in limestone does not react rapidly with sulfur dioxide; the sorbent probably depends for its desulfurizing action on the extent of calcination which occurs at or near the bed air inlet. Both sulfur dioxide and carbon dioxide will then compete for the calcined lime.

Although Exxon (2) has been successful in using limestone as a sorbent at pressure, it is uncertain if there are high-pressure and low-temperature limits to its use. Dolomite, on the other hand, is normally reactive in the half-calcined state and may be used as sorbent, irrespective of the carbon dioxide pressure in the system. The calcium-to-sulfur molar ratio fed to the bed, however, will be higher with half-calcined than with fully-calcined dolomite.

For those stones for which thermogravimetric (TG) data are available, an estimate of performance can be made, given the proposed operating conditions. For any proposed system it is also possible to carry out TG tests on a particular stone to estimate its suitability as a sorbent under the system operating conditions. Data on the kinetics of dolomite and limestone sulfation covering the major variations in stone type encountered in the eastern United States are desirable.

Attrition Resistance

The term attrition is used to cover all aspects of loss of sorbent from the fluidized bed by elutriation, irrespective of the mechanism. Decrepitation, thermal shock shattering, bursting apart during calcination, fragmentation during sulfation, abrasion between sorbent and ash and sorbent and refractory or internals, distributor jet impingement, and sorbent circulation through ducts may all contribute to loss of sorbent from the bed. It is impossible to establish criteria for attrition resistance which must be met without reference to the system design, since the system design itself may take into account the attrition loss expected from available sorbents. In general, however, a minimal amount of attrition loss is desirable, if only to reduce the load on dust removal equipment.

Additionally, for regenerative systems where long cumulative residence times in the desulfurizing system are required, the rate of calcium makeup feed to the bed may be governed by attrition losses. At the other end of the scale, continuous recirculation of fines assists in attaining higher stone utilization in once-through systems, or in systems with a high calcium-to-sulfur molar feed rate. The prediction of attrition losses a priori for any particular stone is not yet possible.

Laboratory tests can give a relative listing of the expected order of attrition losses from a series of stones. The evidence, however, is that suitable adjustment of the process variables may alter attrition loss sufficiently to accommodate a particular stone to a particular process. For example, Pope, Evans and Robbins (PER) (3) successfully modified initial calcination conditions in a fluidized bed combustor to permit operation with a local, impure limestone.

Another approach is to harden the stone sorbent to increase the number of cycles for which the sorbent can be circulated between desulfurization (under reducing conditions) and regenerator, permitting a low sulfur differential between the two vessels. Experiments to compare attrition losses for a range of dolomites and limestones of different grain structure for empirical evaluation of stone sorbents are necessary to assist in the screening of candidate stones.

Trace Element Emission Characteristics

The possibility of the emission of trace elements from the sorbent into the effluent gas from the fluidized bed combustor raises evironmental and corrosion questions.

The major concern is that sodium and potassium liberated into the gas stream will form liquid deposits on turbine components, thereby inducing hot corrosion. Preliminary results indicate that a very small fraction of the alkali-metal content is liberated from dolomites during fluidized bed combustion. The major alkali-metal species emitted are potassium compounds; the best approach is to use the dolomite with the lowest alkali-metal content. Analysis of a range of dolomites from Ohio, Indiana, Illinois, and Michigan indicates that the potassium content varies from 100 to 6,000 ppm (by weight), while the sodium content varies from 150 to 350 ppm (by weight). Occasional large deviations from these values may be encountered: for example, a Bahamian aragonite dredged from the sea contains 4,000 ppm sodium. To date, no correlation of alkali-metal content with geological formation has been found.

The sparse data available show that the alkali content may vary significantly through a series of rock strata, and within one stratum from quarry to quarry. The best recommendation is thought to be to choose as sorbent from the available dolomite or limestone that which is lowest in potassium content.

Suitability of Spent Sorbent for Final Processing for Disposal

Any final processing of the spent sorbent required before disposal is probably a local problem, dependent on the particular fluidized bed combustion system operated, on regulations governing disposal of solid wastes, or on local marketing opportunities. The major influence these factors may have is in requiring a choice between limestone or dolomite. Investigation of this aspect of stone selection is dependent on the production of tonnage quantities of spent sorbent so that its properties may be characterized.

Economic Availability of the Stone

The cost of dolomite in the eastern and midwestern states was in the range of $2.30 to $5.00/Mg (ton) in late 1974. Significant transport costs are likely if the sorbent must be hauled for long distances. For this reason, operation with local sources of stones is preferable. While systems studies reveal the impact of sorbent cost on plant economics, additional work to determine the design conditions needed to operate (desulfurize) with a wide variety of stone types is equally important. Successful operation of the Rivesville plant with a local, impure limestone indicates the potential for this approach. TG work to modify the properties of the stone sorbents by special calcination treatment in order to develop suitable stone porosity for desulfurization was successful with pure and impure dolomites and with a pure limestone.

ALTERNATIVE SORBENTS

Assessment of Properties

Because of the low cost and widespread geographical availability of calcium carbonate as limestone or dolomite rocks and its excellent performance as a sorbent in preventing sulfur dioxide emission, a compelling reason must exist before an alternative sorbent is considered as a substitute. Apart from cases where an alternative to calcium carbonate might be considered because of local availability, the most general ground for assessing alternative sorbents is that of regenerability.

Calcium sulfate is extremely stable and requires the expenditure of considerable energy before it will release sulfur, either as sulfur dioxide or as hydrogen sulfide, in concentrated form. The desirability of regeneration rests on two factors, the recovery of sulfur as a valuable resource and a reduction in both the material requirements for sulfur sorption and the quantities of spent sorbent solid which are produced as a by-product of the sulfur removal process.

The ideal alternative sorbent is one which is mechanically resistant to attrition in the solids circulation system of the fluidized bed process; is an efficient sulfur getter over the range of coal combustion conditions in the fluidized bed process; can be regenerated under mild reducing conditions (or by thermal decomposition alone); and yields a solid oxide which retains the capacity and kinetic activity of the original sorbent on repeated recycling. In addition, it should contribute no fine particulate matter or trace elements to the effluent gases and should trap these materials as they are released from coal and coal ash during combustion. A large number of metal oxides have been screened for their suitability in removing sulfur dioxide from flue gases. The thermodynamic data tabulated

for the stability of metal sulfates rule out most of these oxides for consideration as sorbents in the fluidized bed combustion process.

The free-energy diagram for metal sulfate stability shown by Bartlett (4) indicates that at 727°C (1341°F) the order of thermal stability of the common
metal sulfates is:

(1) Potassium sulfate   (5) Nickel sulfate
(2) Sodium sulfate   (6) Copper sulfate
(3) Calcium sulfate   (7) Aluminum sulfate
(4) Manganese sulfate

Calcium sulfate requires severe treatment (i.e. high temperature) in reducing gas to
release the sulfur, indicating that more stable alkali-metal sulfates can be excluded.

Evaluation of Copper-Containing Compounds

A special alumina-based copper oxide catalyst, Nalco 471, has been evaluated
as a potential sulfur removal sorbent. Both the thermodynamics and kinetics
of the reaction of sulfur dioxide with the catalyst were examined.

Thermodynamic Evaluation: Copper sulfate ($CuSO_4$) decomposes thermally in
two stages as shown by the equations:

$$(1) \qquad 2\,CuSO_4 \rightleftharpoons CuO \cdot CuSO_4 + SO_2 + {}^1\!/_2 O_2$$

$$(2) \qquad CuO \cdot CuSO_4 \rightleftharpoons 2\,CuO + SO_2 + {}^1\!/_2 O_2$$

At a given temperature in the range of interest for fluidized bed combustion
the equilibrium pressure of sulfur dioxide is lower for reaction (2) than for
reaction (1). Therefore, reaction (2) should be considered as the process which
will thermodynamically limit the sulfur retention in a fluidized bed of copper
oxide. The degree of dispersion of copper oxide on the alumina base, however,
may either prevent formation of the oxysulfate or, indeed, stabilize it. Mixtures
of sodium sulfate and copper oxide are more effective sulfur dioxide sorbents
than copper oxide alone; the sulfation reaction is faster than with copper oxide;
and thermal decomposition requires heating to a higher temperature (5).

The limitations which reaction (2) places on SO_2 removal are shown in Figure 5.2
for: (A) the 101.3 kPa (1 atm) combustor, (B) the 1013 kPa (10 atm) combustor
with 10% excess air, (C) the 1520 kPa (15 atm) combustor with 10% excess air,
and (D) the 1520 kPa (15 atm) adiabatic combustor. Table 5.1 shows the assumed exit gas conditions. Calculations indicate that sulfur retention would fall off
from more than 90 to 20% in the range 627° to 727°C (1161° to 1341°F).

TABLE 5.1: EXIT GAS CONDITIONS FOR FLUID BED COMBUSTORS

System	P_{SO_2} with No Sorbent (atm)	P_{O_2} (atm)
(A) 101.3 kPa (1 atm) FB, 10% excess air	0.005	0.04
(B) 1013 kPa (10 atm), 10% excess air	0.05	0.17
(C) 1520 kPa (15 atm), 10% excess air	0.075	0.255
(D) Adiabatic combustor, 1520 kPa (15 atm)	0.0188	2.265

Source: PB 246 116

FIGURE 5.2: MAXIMUM SO$_2$ RETENTION IN FLUIDIZED BEDS OF CuO·CuSO$_4$ (Thermodynamic Limit)

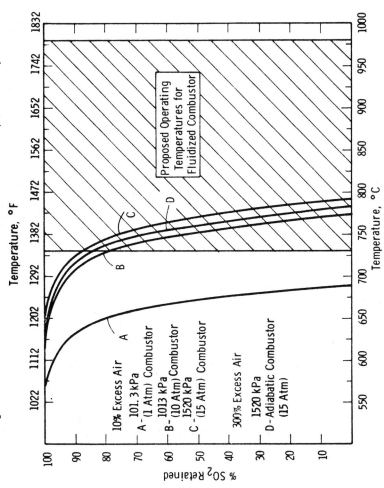

Source: PB 246 116

Experimental Results: TG experiments were carried out at atmospheric pressure using the copper-impregnated alumina catalyst and an oxide prepared by thermal decomposition of copper sulfate pentahydrate.

The copper sulfate pentahydrate decomposed to yield a 32.1% weight loss for dehydration (theoretical value = 32.3%) and a 36.1% weight loss for decomposition of sulfate to oxide (theoretical value = 36.1%). The oxide was then exposed to 12% sulfur dioxide/nitrogen/4% oxygen and the temperature varied in the range 600° to 800°C (1112° to 1472°F). Equilibrium temperatures were noted at the transition from weight gain to weight loss.

As expected, two equilibrium temperatures were noted, and the temperatures were in relatively good agreement with those predicted from the experimental thermodynamic data in the literature (6).

The Nalco 471 catalyst, impregnated with copper oxide was then tested in the same sulfur dioxide/nitrogen/oxygen mixture. For runs with two samples, only one equilibrium temperature was found at 684 ± 5°C (1263 ± 41°F). This apparently anomalous behavior results from the reaction:

$$Al_2(SO_4)_3 \rightleftharpoons Al_2O_3 + 3SO_2 + \tfrac{3}{2}O_2$$

Calculation of the temperature for equilibrium at the experimental gas pressures, using the thermodynamic tabulation of Stern and Weise (7) gave 680.5°C (1257°F). In addition, the total quantity of sulfur trioxide (SO_3) reacted was about three times stoichiometric for formation of copper sulfate (on a 5 weight percent copper basis). It was decided to reverse the experimental procedure and start the absorption/temperature run at high temperatures, cool the solid, and determine the temperature for formation of either copper sulfate or copper oxysulfate free from interference from the alumina.

Sorbent Studies: Pressurized sulfation runs (1013 kPa/10 atm) were carried out on the Nalco catalyst. A trace reaction was noted at 900°C (1652°F). On cooling the sample in the gas flow (0.5% sulfur dioxide; 4% oxygen in nitrogen), the sorbent gained weight at the temperature where aluminum sulfate becomes thermodynamically stable. The sample lost weight on recycling through this temperature [650°C (1202°F)]. A blank run on colloidal alumina showed a trace reaction at 900°C (1652°F), similar to that observed for the Nalco catalyst, a weight gain corresponding to 1% of the original sample weight.

It was concluded that the copper-impregnated catalyst is not a suitable sorbent for sulfur dioxide removal in fluidized bed combustion: the sulfur dioxide adsorption step takes place below the practical temperature range for operation of the fluidized bed combustor.

General Assessment

The chief interest in alternative sorbents lies in finding a sorbent which can undergo multiple regeneration. For this reason an analysis of the minimum acceptable performance of a sorbent should be carried out. This analysis requires definition of:

(1) A range of acceptable hydrogen sulfide or sulfur dioxide concentrations produced in the regenerator;

(2)   Acceptable fuel consumption in the regenerator;
(3)   Minimum sulfur loading on the sorbent and acceptable stone recirculation rates;
(4)   Thermodynamic screening of the sorbents in light of (1) and (2) and
(5)   Kinetic tests for sorbent activity in the light of (4).

The criteria developed would be system dependent: the atmospheric pressure, pressurized boiler, and adiabatic combustor cases would each be treated separately.

SULFATION OF LIMESTONE AND DOLOMITE

Thermogravimetric Studies and Experiments

Thermogravimetric (TG) studies of the sulfation of calcium-based sorbents show a close correspondence with fluidized bed work. In particular, they show that calcium utilization in the sulfation of dolomite at pressure is highly dependent on the conditions of carbon dioxide partial pressure which prevail during stone calcination. Studies have been extended to consider both the sulfation of limestone and dolomite at atmospheric pressure, and the sulfation of half-calcined dolomite at pressure.

Laboratory studies on the sulfation of lime and calcined dolomite indicate:

Sulfation is first order with respect to sulfur dioxide.
Sulfation is zero order with respect to oxygen.
The rate of reaction increases with temperature. Various values
of activation energy have been derived, but there is no unambiguous determination of activation energy.
Reactivity depends upon the pore volume formed during calcination:
Small pores give a high initial reaction rate but, if the particles
are large, a low overall capacity results.
Large pores give lower rates but increased capacities.
For small particles, capacity is determined by the pore volume
available for product accumulation.
Other properties which influence the pore structure formed on
calcination may serve as general indices to the capacity, in other words, temperature and conditions of calcination, and sodium content of the stone.
The distribution of sulfur through a sulfated stone depends on
the stone: a coarse limestone shows total permeation of the stone by sulfate, while Iceland spar forms a rim of sulfate.

Experiments were carried out to determine:

The effect of calcination conditions on the sulfation of limestone
and dolomite;
The effect of temperature on limestone sulfation;
The effect of pressure and temperature on limestone sulfation and
The effect of temperature on the reaction between sulfur dioxide
and calcium oxide.

The materials used in the experiments and their composition are listed in Table 5.2.

TABLE 5.2: SORBENTS USED TO STUDY THE SULFATION REACTION

Sorbent	Ca	Mg	Ignition wt. loss	Principal impurities
Limestone 1359 Stephens City, Va.	38.4	.04	43.4	Silica
Tymochtee Dolomite Huntsville, Oh.	20.5	11.9	44.4	Silica, pyrites, alumino silicates
Glasshouse Dolomite (Dolomite 1337) Gibsonberg, Oh.	21.5	12.5	47.7	Silica, pyrites, alumino silicates
Salamonie Dolomite Portland, Ind.	21.7	12.9	47.9	
Canaan Dolomite New Canaan, Conn.	22.2	12.8	46.0	Traces of amphibole

Source: PB 246 116

Effect of Calcination Conditions

Experimental runs aimed at determining the effect of calcination conditions on the sulfation of limestone 1359 are listed in Table 5.3.

In the first set of experiments, at 1013 kPa (10 atm) pressure, Experiments 1 through 4, the effect of carbon dioxide partial pressure during calcination was probed. By calcining the limestone in nitrogen, either rapidly at 900°C (1814°F), or slowly in the temperature range 680° to 870°C (1256° to 1598°F), a lime was formed which was relatively inert and ceased to sulfate at a rapid rate after 14% utilization of the calcium. When calcination was retarded, how-ever, by maintaining 101.3 kPa (1 atm) of carbon dioxide over the solid and heating it to 930°C (1706°F), a more active lime was produced which yielded 32 and 37% calcium utilization in successive experiments. A second set of runs at atmospheric pressure, Experiments 5 through 8, demonstrated that the activa-tion of the lime is not due to pressurized operation.

Since the carbon dioxide partial pressure during calcination controls the activity of the product lime during sulfation, even at atmospheric pressure, tests were also run using dolomite 1337 as sorbent at atmospheric pressure. It was found that dolomite could be activated at atmospheric pressure by calcining it under a high partial pressure of carbon dioxide.

At 1013 kPa (10.0 atm) and 101.3 kPa (1 atm), the two pressures studied, dolomite is a superior sorbent, both on the basis of the weight of raw sorbent used and on the basis of calcium utilization. The possibility remains, however, of greatly increasing the capacity of the limestone; however, over 80% of the calcium in the dolomite is sulfated, leaving only a small margin for further improvement.

Fluidized Bed Calcine

A set of experiments was run on the sulfation of lime calcined in the 50 mm fluidized bed unit. The calcined sample was obtained in a test on attrition

TABLE 5.3: LIMESTONE 1359 SULFATION RUNS

Experiment Number	Pressure, kPa (atm)	Particle size, μm	Calcination			Sulfation	
			Temperature, °C (°F)	Atmosphere	Time/min.	Temperature, °C (°F)	Utilization, % Ca
1	1013 (10.0)	420-500	900 (1652)	N_2	4.0	871 (1600)	14.0
2	1013 (10.0)	420-500	680-870 (1256-1598)	N_2	~ 12.0	871 (1600)	14.0
3	1013 (10.0)	420-500	930 (1706)	10% CO_2/N_2	8.0	871 (1600)	32.0
4	1013 (10.0)	420-500	930 (1706)	10% CO_2/N_2	8.0	871 (1600)	~ 37.0
5	101.3 (1.0)	420-500	900 (1652)	N_2	2.5	871 (1600)	9.0
6	101.3 (1.0)	420-500	900 (1652)	30% CO_2 in N_2	5.0	871 (1600)	14.0
7	101.3 (1.0)	420-500	871 (1600)	55-30% CO_2 in N_2	80.0	871 (1600)	34.5
8	101.3 (1.0)	420-500	900 (1652)	60% CO_2 in N_2	30.0	871 (1600)	42.0
9	101.3 (1.0)	420-500	954 (1749)	15% CO_2 in N_2	~ 1.2	954 (1749)	12.0
10	101.3 (1.0)	420-500	843 (1549)	15% CO_2 in N_2	34.0	843 (1549)	14.0
11	101.3 (1.0)	420-500	899 (1650)	15% CO_2 in N_2	3.0	900 (1652)	11.0

Source: PB 246 116

behavior of limestone 1359. Since the sample was allowed to calcine in a stream of nitrogen in the temperature range 650° to 750°C (1202° to 1382°C), the stone produced was relatively inactive. The extent of sulfation achieved in TG experiments from 101.3 to 1013 kPa (1 to 10 atm) lay in the range 13.5 to 17.5%, as shown in Table 5.4. These experiments confirmed that calcination at low partial pressures of carbon dioxide does produce an inactive stone.

TABLE 5.4: TG SULFATION OF LIMESTONE 1359 CALCINED IN THE 50 mm FLUIDIZED BED*

Experiment Number	$P/(1.03 \times 10^5 \text{ N/m}^2)$	% CaO utilization	
		After 5 min	After 1 hr
12	1.0	11.7	17.5
13	10.0	13.9	17.8
14	10.0	11.9	15.1
15	5.0	10.7	13.5

*Initial particle size 1,000 to 1,420 μm; calcined in nitrogen at 650° to 750°C (1202° to 1382°F); sulfation at 5,000 ppm SO_2, 4% O_2 in N_2 at 870°C (1598°F).

Source: PB 246 116

Temperature Effect

Studies on the sulfation of limestone and dolomite have demonstrated that the effect of temperature on the reaction is a complex phenomenon. Activation energies obtained at low calcium utilization show values in the range of 5 to 15 kcal-mol, 20 to 62 kJ, indicative of a mixture of mass transport and chemical reaction control. In fluidized beds, however, where sulfur dioxide removal is normally carried out at 30 to 40% utilization, a maximum in the extent of sulfur dioxide removal at a fixed calcium utilization is observed at a temperature of about 843°C (1550°F) at atmospheric pressure, implying that the rate of the overall reaction decreases above 843°C (1550°F).

Calcination of samples was effected at 930°C (1706°F) in 60% carbon dioxide in nitrogen, and the samples were then sulfated at temperatures from 750° to 950°C (1382° to 1742°F) at 50°C (122°F) intervals, in random order. The results are shown in Figure 5.3 and indicate that although the initial rates were scarcely distinguishable, the course of reaction was different for each temperature after 20% utilization of the calcium. The importance of these differences is evident when the extent of reaction after a fixed time interval, one hour, is plotted as a function of sulfating temperature in Figure 5.4. The utilization of the stone peaks at a maximum value of 43% at about 860°C (1580°F), reproducing the sulfur dioxide retention efficiency pattern observed in fluidized bed, results. The maximum rate of sulfation noted was 3.99 mg/min.

FIGURE 5.3: THE EFFECT OF TEMPERATURE ON LIMESTONE SULFATION

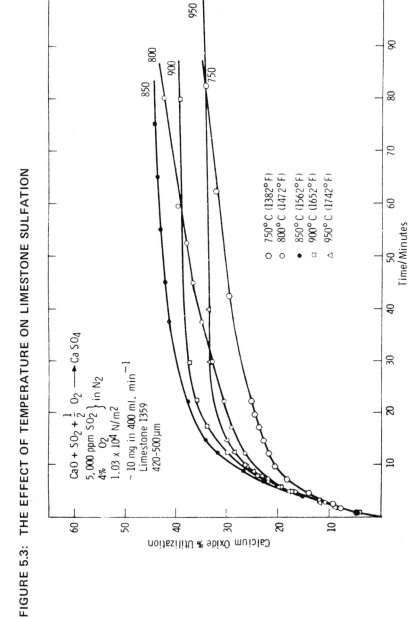

FIGURE 5.4: THE EFFECT OF TEMPERATURE ON CaO UTILIZATION
IN SULFATION

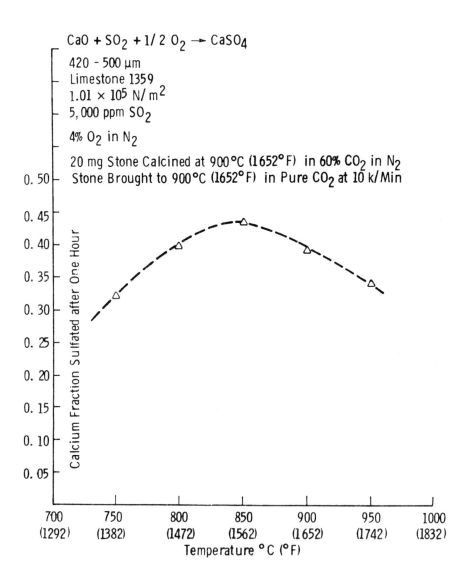

Source: PB 246 116

Sulfation of Dolomite

The activation process of calcining under a high partial pressure of carbon dioxide yields a dolomite with greater sulfur dioxide sorption capacity at a fast rate of reaction. The effect of an increase in particle size from 420-500 μm to 1,000 μm was studied. Dolomite 1337 of 1,000 μm particle size was calcined in 10% carbon dioxide in nitrogen at 750°C (1382°F) to calcine the magnesium carbonate fraction.

The sample was then heated to 925°C (1695°F) and the carbon dioxide partial pressure dropped to 81.04 kPa (0.8 atm). The sample was calcined and sulfated at 871°C (1600°F) in 0.5% sulfur dioxide in 4% oxygen in nitrogen. A comparison sample, was similarly sulfated after calcination under nitrogen by heating to 880°C (1616°F). Comparison of the rates of reaction showed that the calcination treatment in carbon dioxide had increased the stone capacity. If 2% calcium/minute is taken as the criterion for an acceptably fast rate of reaction with sulfur dioxide under the given experimental conditions, then the rate of reaction fell below this value at 50% utilization in the first sample and 35% utilization in the second.

This can be regarded as an improvement of the calcium/sulfur molar feed ratio from 2.6 to 1.8 at a mean particle size of 1,100 μm. The activation process for increasing the utilization of dolomite can be used with large particles of sorbent.

Sulfation of Half-Calcined Dolomite

As Figure 5.1 illustrates, half-calcined dolomite is stable over much of the operating range of the low excess-air fluidized bed combustor. The stability line shown in the figure is for the exit gas carbon dioxide composition; at the bed entrance, where the fluidized gas is mainly air, some calcination will take place, yielding fully-calcined dolomite, and sulfur dioxide and carbon dioxide will compete for the calcium oxide generated.

It has been shown that half-calcined dolomite is a reactive sorbent for sulfur dioxide removal. The runs, however, used larger samples than are optimal, and the kinetics of the reaction can not be assessed relative to the reactivity of calcined dolomite.

Three experiments were carried out on the sulfation of half-calcined dolomite 1337. The first two experiments were tests with 1,680 to 2,000 μm diameter particles.

In the first experiment the temperature was cycled between 850° and 690°C (1562° and 1274°F) as the reaction proceeded. After 23 minutes 33% utilization was noted, during which time the sample had been cooled to 690°C (1274°F) and heated back to 820°C (1508°F). In the following experiment isothermal sulfation at 800°C (1472°F) gave 50% utilization in 23 minutes. The decline in reaction was not as abrupt as has been universally noted with calcined dolomite after two hours of reaction; 87% utilization was noted in the isothermal experiment (Figure 5.5).

A third experiment on (420-500 μm) 1337 half-calcined dolomite showed greater initial reactivity, 61% utilization in 23 minutes. The subsequent activity, however, was slow, yielding 68% utilization after two hours. The reactivity of half-calcined dolomite appears to be intermediate between nitrogen-calcined dolomite and dolomite which has been calcined under a high partial pressure of carbon dioxide relative to the equilibrium pressure. In addition, the slowing down of reaction is less abrupt than with calcined dolomite or limestone. This latter fact means that increased gas residence times in the fluidized bed should improve (lower) the calcium/sulfur molar ratios required for any given sulfur removal criterion.

FIGURE 5.5: SULFATION OF HALF-CALCINED DOLOMITE

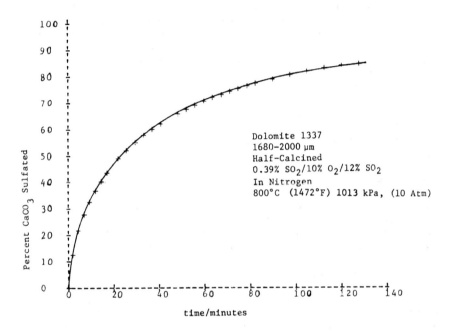

Source: PB 246 116

Assessment of Thermogravimetric Experimental Studies

Additional studies of temperature effects, calcination of limestone and dolomite, rate of limestone calcination and rate of dolomite calcination have led to the following conclusions.

The sulfation of limestone proceeds to higher calcium utilization at a fast rate of reaction at atmospheric pressure and at 1013 kPa (10 atm) if the pressure of carbon dioxide (CO_2) has been high (approximately $0.6 \times P_{eq}$) during the limestone calcination step (where P_{eq} is the equilibrium pressure of carbon dioxide over calcium carbonate at the reaction temperature).

The effect of temperature on sulfur dioxide (SO_2) retention in fluidized beds of limestone is governed by the above calcination effect and by the effect of temperature on the reaction of lime with sulfur dioxide and oxygen, not primarily by the oxidation/re- duction cycle in the fluidized bed.

Pressurized reaction may require the use of half-calcined dolomite or uncalcined limestone. Most half-calcined dolomites are ex- cellent sulfur dioxide sorbents, intermediate in reactivity between dolomite calcined under low carbon dioxide pressure and dolomite calcined at high carbon dioxide pressure. Uncalcined limestone is an extremely poor sorbent for sulfur dioxide; calcination and recarbonation does not improve its sorbent properties.

SORBENT UTILIZATION AND SULFUR REMOVAL EFFICIENCY

Study Objectives

Thermogravimetric (TG) studies on the sulfation of limestone and dolomite are performed to obtain design parameters for fluidized bed desulfurization systems. As stated, it is postulated that (1) the rate of sulfur sorption by particles of limestone or dolomite by the overall reaction

$$CaO + SO_2 + \frac{1}{2}O_2 \longrightarrow CaSO_4$$

and (2) the total capacity of the stone to absorb sulfur are primary factors in determining the usefulness of a particular sorbent, granted the thermodynamic conditions for sulfur dioxide (SO_2) removal are favorable. TG studies can there- fore be assumed, a priori, to provide an estimate of the relative impact of particle size, temperature, pressure, gas composition, and choice of sorbent on desulfur- ization in a fluidized bed. The TG data should, therefore, permit a qualitative assessment of the effect of these parameters on the design and operation of fluidized beds.

Reaction Kinetics and Model Studies

There is a need for data under closer simulation of fluidized bed operating con- ditions (1013 to 1520 kPa; 10 to 15 atm pressure). In a TVA study including ki- netic measurements on limestone 1359, the maximum degree of sulfation observed was 18% (8). Using the same type of stone, Exxon workers found that 32% of the stone was sulfated in fluidized beds before significant escape of sulfur dioxide from the bed was observed (9).

Similar particle sizes and oxygen pressures were used at atmospheric pressure in both tests. The results are significant because Borgwardt (10) has established that the limit on fast sulfation kinetics is determined by the space available in the stone pores to accommodate the bulky sulfate ion. This limit should not depend on whether the reaction takes place in a TG apparatus or a fluidized bed. Application of kinetic studies to projections of fluidized bed desulfurization has been carried out by several groups. Conclusions drawn from these studies and models developed for predicting the effect of calcium/sulfur (Ca/S) ratio on desulfurization; temperature on desulfurization; bed height and gas velocity on desulfurization; and pressure on desulfurization are as follows.

Ca/S mol ratios of approximately 1/1 can achieve 90% desulfurization if the dolomite is not calcined under low partial pressures of carbon dioxide. Pressurized operation, therefore, which
maintains a high P_{CO_2} in the bed should tend towards the 1/1
calcium/sulfur molar ratio.

Half-calcined dolomite should be used as a sorbent in the range
intermediate between 1/1 and 2.5/1 calcium/sulfur molar feed
ratios. Improvement in desulfurization can be affected by modest
increases in bed height or gas residence time.

More advanced models should consider the effect of particle size
distribution; the effect of high-temperature operation above 1750°C
(3182°F); and the effect of grain structure of the raw dolomite
on its capability to be calcined so that it can be used at low
calcium/sulfur molar ratios.

REFERENCES

(1) G.P. Curran, C.E. Fink and E. Gorin, *CO₂ Acceptor Gasification Process in Fuel
     Gasification.* Consolidation Coal Co. Advances in Chemistry Series 69. American
     Chemical Society, Washington, D.C. 1967, page 141.

(2) R.C. Hoke, L.A. Ruth and H. Shaw, *Combustion and Desulfurization of Coal in a
     Fluidized Bed of Limestone.* Exxon Research and Engineering Co., Linden, N.J.
     (Presented at IEEE-ASME Joint Power Generation Conference, Miami Beach, Fla.
     Sept. 15-19, 1974.)

(3) J.E. Mesko, S. Ehrlich, and R.A. Gamble. *Multicell Fluidized-Bed Boiler Design Con-
     struction and Test Program.* Office of Coal Research. Pope, Evans and Robbins,
     New York, N.Y. NTIS PB 236 254, August 1974.

(4) R.W. Bartlett, *Sulfation Kinetics in SO₂ Absorption from Stack Gases.* Environmental
     Protection Agency. Stanford University, Palo Alto, California. Grant No. AP00876.
     June 1972.

(5) F.T. Bumazhmov, "Sulfation of Copper Oxide by Mixtures of Air and Sulfur Dioxide".
     Izv. Vyssh. Ucheb. Zaved., Tsvet. Met. 22-6, 1973. (*Chemical Abstracts* 79, 55489C).

(6) T.R. Ingraham, "Thermodynamics of the Thermal Decomposition of Cupric Sulfate
     and Cupric Oxysulfate". *Trans. Met. AIME* 233: 359, 1965.

(7) K.H. Stern and E.L. Weise, *High Temperature Properties and Decomposition of
     Inorganic Salts. Part 1 - Sulfates.* NSRDS-NBS7. National Bureau of Standards,
     Washington, D.C., 1966.

(8) J.D. Hatfield, Y.K. Kim, R.C. Mullins, and G.H. McClellan, *Investigation of the Reactivi-
     ties of Limestone to Remove Sulfur Dioxide from Flue Gas.* Air Pollution Control
     Office, Tennessee Valley Authority, 1971.

(9) G.A. Hammons and A. Skopp, *A Regenerative Limestone Process for Fluidized Bed
     Coal Combustion and Desulfurization.* Environmental Protection Agency. Report
     to A.P.C.O. Exxon Research and Engineering Co., Linden, N.J., February 1971.

(10) R.H. Borgwardt and R.D. Harvey, *Env. Sci. Technol.* 6: 350, 1972.

SPENT STONE DISPOSAL

The material in this chapter was excerpted from PB 246 116. For a complete bibliography see page 263.

FACTORS AFFECTING SPENT STONE DISPOSITION

Among the factors that will affect the disposition of the spent sorbent are, for example, the quantity of spent sorbent, its chemical characteristics, regulations, geographical location, and the size of the market for respective applications. A summary of the general process options for disposition of the spent sorbent is presented in Figure 6.1.

Quantity

Representative quantities of spent material are illustrated in Table 6.1 for application in fluidized bed combustion systems. Typical quantities of spent sorbent projected for disposal from a 500 MW power plant burning a 3 wt. % sulfur fuel with 95% removal would range from approximately 41 Mg/hr (45 tons/hr for a once-through system using dolomite to approximately 19 Mg/hr (21 tons/hr) for a regenerative system utilizing a calcium/sulfur makeup rate of 0.75. Specific quantities will depend on operating conditions and sorbent characteristics.

Chemical Characteristics

The chemical characteristics of the spent sorbent will depend on a number of operating and design parameters, for example: the operating conditions, the specific stone utilized in the process, and the location of the effluent stream from the process (for example, in a regenerative process the stone for disposal can be removed before or after the regeneration process).

The primary constituents are calcium sulfate, calcium oxide (or calcium carbonate), and magnesium oxide when dolomite is used; and calcium sulfate and calcium oxide (or calcium carbonate) when limestone is used.

105

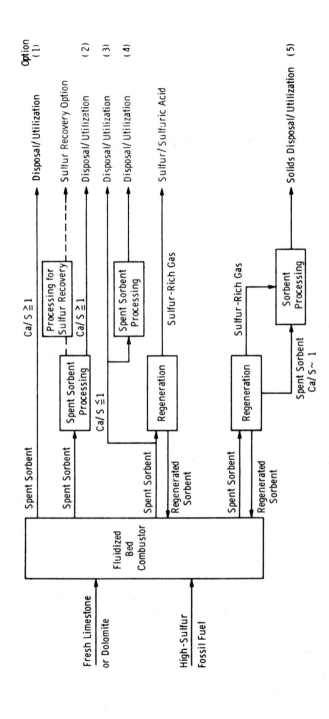

FIGURE 6.1: SULFUR REMOVAL SYSTEM PROCESS ALTERNATIVES

Source: PB 246 116

TABLE 6.1: SPENT STONE DISPOSITION

	1970	1975	1985	2000
Total Utility Consumption				
Coal, Tg (10^6 ton)	296 (326)	369 (407)	485 (535)	708 (780)
Sulfur[a], Tg (10^6 ton)	8.9 (9.8)	11 (12.2)	14.6 (16.1)	21.2 (23.4)
Fluidized Bed Installation Factor[b]	0.0	0.0	0.09 (0.10)	0.41 (0.45)
Spent Dolomite[c], Tg (10^6 ton)				
• Once-through	0.0	0.0	9.5 (10.5)	62.6 (69.0)
• Regenerative	0.0	0.0	4.9 (5.4)	31.8 (35.0)
Landfill, (acre-ft)				
• Once-through	0.0	0.0	4,560	30,000
• Regenerative	0.0	0.0	2,350	15,200

[a]Sulfur in coal at 3 percent.

[b]Ratio of electrical energy produced in coal-fired plants with fluid-bed units to electrical energy produced in coal-fired plants with other combustion units.

[c]Assumes dolomite Ca/S molar ratio of 1.5 for once-through and 0.75 for regeneration.

Source: PB 246 116

The amount of magnesium oxide present is determined by the magnesium carbonate ($MgCO_3$) composition of the fresh stone. The distribution of the calcium-based compounds in the spent sorbent is summarized in Table 6.2.

TABLE 6.2: TYPICAL COMPOSITION OF CALCIUM COMPOUNDS IN SPENT SORBENT

	CaO or $CaCO_3$ [a] mole %	$CaSO_4$ mole %
Once-through		
Dolomite	10	90
Limestone	50	50
Regenerative		
Dolomite/limestone	10-70	30-90 [b]

[a] The oxide or carbonate form is determined by the fluidized bed combustor operating temperature, pressure, and excess air.
[b] The $CaSO_4$ composition is determined by the sulfur loading for the sorbent to the regenerator.

Source: PB 246 116

The solubility of calcium and magnesium compounds is of particular importance in determining chemical activity. A compilation of readily available solubility data on calcium and magnesium compounds was made to aid the study. The oxide form, as well as carbonates, phosphates, and silicates, of magnesium have minimal environmental impact. Magnesium sulfites and sulfates would not be inert. The concentration of these compounds, however, appears negligible.

This raises the possibility, however, of achieving a separation of dolomitic sorbents into calcium and magnesium compounds which separately might have market potential. One problem with phosphate forms is that phosphoric acid is relatively expensive, and there does not appear to be an inexpensive source of the phosphate ion for spent stone processing.

Regulations

An important aspect in the development of a viable disposal process is the constraint of existing regulations relative to environmental impact. Each site will require examination of both local and federal regulations.

Recent solid waste disposal laws recognize that the problems of waste disposal have become national in scope and require federal action to promote the development of solid waste management and resource recovery systems and, also, an overall national research and development program. Legal documents and technical investigations which have been reported to carry out the provisions of these acts will provide background information and criteria for disposition of spent sorbent from the fluidized bed combustion process.

Location and Market

The geographical location of each plant determines the sorbent utilized and the disposal options. Limestones and/or dolomites vary geographically and factors affecting sorbent selection, etc. have already been discussed. The question of market compatibility with by-products must be considered for each case. Illustrations of possible applications are presented in the following sections. One aspect of the market picture is to assess the total market available. For example, the domestic consumption of magnesium compounds is of interest.

Recent data on the domestic usage of magnesium compounds indicate the two largest uses were high-temperature refractory applications—dead-burned dolomite and refractory magnesia. This is not a likely market for the spent stone sorbent from the fluidized bed combustion process in the nonregenerative versions. The effluent sorbent would in these cases have a high sulfate content which could not be retained at refractory use-temperatures. The regenerative versions with low calcium sulfate loading, however, might be able to supply stone to this market.

At 330 on-stream days per year, a 500 MW plant would produce on the order of 25,000 Mg (27,557 tons) of magnesium per year in various forms (fully-calcined dolomite, half-calcined dolomite, and so on) in the regenerative process, and twice that quantity in the nonregenerative versions. It appears that the potential market for refractory forms of magnesia and dolomite could be large enough to accommodate the output from a substantial number of 500 MW plants.

If the spent sorbent cannot be used as dead-burned dolomite, the market size shrinks substantially. In this case new, large-scale uses would have to be developed. Other potential markets exist for the spent sorbent. A comprehensive assessment of the alternatives is required, followed by tests with actual spent sorbent for the preferred options.

DISPOSITION OF UNPROCESSED SPENT SORBENT

Preliminary leaching experiments and activity tests of unprocessed spent sorbent (represented by options 1 and 3 in Figure 6.1) indicate

   (1)   In both a leaching-time experiment and a stone-loading experiment calcium and sulfate dissolution plateaued at concentrations limited to the calcium sulfate solubility.

   (2)   The equilibrium calcium and sulfate concentrations were high, exceeding the water quality criteria. Since calcium sulfate occurs abundantly in nature as gypsum, leachates induced from a natural gypsum offer a good reference for its calcium and sulfate concentrations. Iowa ground gypsum No. 114 was selected to undergo parallel leaching tests with the waste stone. Results indicated that gypsum leachates contained approximately the same amounts of dissolved calcium and sulfate ions as the test leachates. Both agreed relatively well with the calcium sulfate solubility, and both exceeded the water quality standards, 75 mg/l for calcium and 250 mg/l for sulfate.

(3) There was negligible dissolution of magnesium ions.

(4) Insignificant amounts of heavy metal ions were found in the leachates.

(5) Test leachates were alkaline, with pH = 10.6 to 12.1; however, the run-off leachates showed a gradual decrease in pH with the amount of leachates passing through.

The spent sorbent characteristics (size and composition) may make the stone attractive for utilization. Areas of particular interest include soil stabilization and landfill. In a general sense, soil stabilization includes any treatment of soil whereby it is made more stable. It is well established for conventional soils that such properties as strength, stiffness, compressibility, permeability, workability, susceptibility to frost and sensitivity to changes in moisture content may be altered by various methods. Such methods range from simple compaction to expansive techniques for grouting, drainage, waterproofing, and strengthening of material by thermal means.

Locally available road-bed soils are stabilized commonly by the addition of agents such as Portland cement, lime, and lime/fly ash mixtures. They act by forming cementitious compounds that more or less permanently bond together individual particles or aggregates of soil.

Tests are required to determine what the characteristics of various types of soils are when they are blended with the spent sorbent. Specifically, tests to determine the unconfined compressive strength, the Atterberg limits, and direct and triaxial shear strength of soils blended with various amounts of the waste materials should be performed. These tests would provide sufficient information to enable specification of soil-waste material mixtures which can be used as load-bearing materials in highway foundation and embankment structures. Wherever necessary, these mixtures of soil and waste material may be further blended with conventional cementitious materials such as Type 1 Portland cement.

DISPOSITION OF PROCESSED SPENT SORBENT

The direct disposal or utilization of spent sorbent may not be possible or permitted in all cases. Thus, alternatives for spent stone disposition must be developed to permit utilization of the fluidized bed combustion process. It may also be possible to develop a more attractive use for the spent sorbent through some processing technology.

Several processes are proposed, as summarized in Figure 6.2. The direct disposal options previously discussed and the final disposition of the sulfur in the regenerative processes are also shown. The basic option for dolomite is to operate either half-calcined or fully-calcined. In each of these modes it is possible to operate regeneratively or nonregeneratively.

Further, there are two options on regeneration: a one-step, high-temperature process, and a two-step, low-temperature process. The former leads to sulfur dioxide, which is probably best routed to an acid plant but can be converted to sulfur. The latter releases hydrogen sulfide, which is convertible to elemental

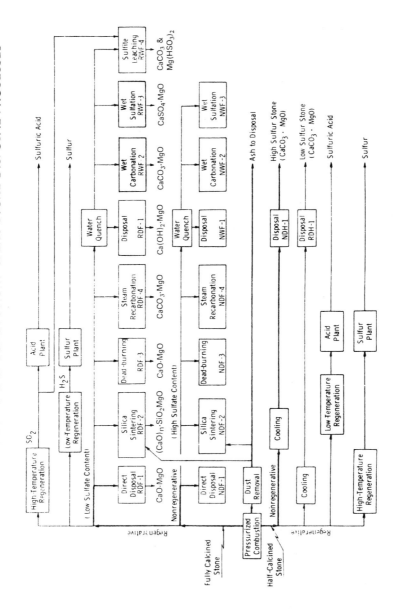

FIGURE 6.2: PRESSURIZED FLUIDIZED BED COMBUSTION: SPENT STONE DISPOSAL PROCESSES

sulfur or acid. In reading the figure, for each of the options on the respective regenerative branches one must include one of the two sulfur processing alternatives. The figure is expressed in terms of dolomite, but similar considerations apply if limestone is used. This figure focuses on the processing of the unsulfated fraction of the stone.

In all cases coal ash is assumed to be taken overhead and removed from the pressurized hot flue gas stream by particulate removal system before sending the gas on to power recovery turbines. The various process options shown on Figure 6.2 are given a letter code for convenience: first letter, R for regenerative and N for nonregenerative (once-through); second letter, D for dry process and W for wet process; and third letter, H for half-calcination (retention of $CaCO_3$) and F for full-calcination (formation of CaO).

Where compositions are shown, only the form of the residual excess calcium compound is given. In every case calcium sulfate is also present as calcium sulfate/magnesium oxide. Brief descriptions of the process options follow.

Disposal: Option NDH-1

This is a simple method for stone disposal, involving only cooling of the stone, since the composition calcium sulfate, calcium carbonate, and inerts is expected to be environmentally inert. While landfill is the most likely end use, it is possible that special cements may be made from it. As a nonregenerative process, all the sulfur captured is removed from the process as calcium sulfate. No sulfur recovery facilities are required.

Disposal: Option RDH-1

This is also a simple stone disposal process, needing only stone cooling. Sulfur processing facilities are needed since it is a regenerative process. The type depends on which regeneration process is used. The final spent stone may have a low sulfur content. This depends on the concentration of calcium sulfate specified for the regeneration process. Spent stone having a low calcium sulfate/magnesium oxide concentration, say 5 wt. %, may be a better candidate for use in the cement industry than the stone from NDH-1.

Direct Disposal: Options RDF-1 and NDF-1

The bulk of the proposed processes deal with fully calcined stone. They fall into two groups according to whether they are regenerative or once-through. It is convenient to discuss the counterparts together.

Both RDF-1 and NDF-1 will have stone cooling. The ability to use this material as landfill will depend on the content and the reactivity of the stone. RDF-1 product might be marketable where an inexpensive alkali is needed, as in municipal waste treatment or acid mine drainage treatment. NDF-1 product has only 33% calcium oxide/magnesium oxide, which probably makes this process less attractive than RDF-1.

Silica Sintering: Options RDF-2 and NDF-2

To take advantage of the free calcium oxide in the spent stone, recovered coal ash, which is expected to be high in silica, is blended with the stone and sintered at temperatures perhaps as high as 1500°C (2732°F). The conditions would be a compromise between what is required for sintering and what must be avoided to prevent loss of sulfur dioxide from the sulfate present. It is likely that a cooling and a grinding step for the spent stone prior to blending will be required. If cements cannot be prepared from the spent stone, it would appear that whatever pozzolanic activity is present would result in a material that would behave as a stable soil in a landfill operation.

Dead Burning: Options RDF-3 and NDF-3

Exposure of materials like calcium oxide and calcium sulfate for sufficient lengths of time to elevated temperatures reduces their surface area and, hence, their general chemical reactivity. Because of the presence of calcium sulfate, the dead-burned product probably could not be used as a refractory. The material, however, is expected to have a minimal impact on the environment if used for landfill because of the reduced activity of calcium and magnesium compounds.

Steam Recarbonation: Options RDF-4 and NDF-4

In both processes the stone is cooled to an intermediate temperature, say 500° to 700°C (932° to 1292°F), pulverized with air in a jet mill, and then contacted with flue gas and steam to convert calcium oxide to calcium carbonate. The steam increases the reaction rate. This provides a method for rendering the sorbent inactive if the calcium oxide activity prohibits direct disposal.

Disposal: Options RWF-1 and NWF-1

These are simple wet processes in which the calcium oxide is merely hydrated to calcium hydroxide to eliminate any problems in handling calcium oxide. The treated stone may be usable as an inexpensive alkali.

Wet Carbonation: Options RWF-2 and NWF-2

Spent stone is quenched with water and subjected to a three-stage carbonation using recycled flue gas. The object is to convert the free calcium oxide to the less active carbonate form to facilitate disposal.

Wet Sulfation: Options RWF-3 and NWF-3

An alternative to carbonation is to use sulfuric acid to convert the free calcium oxide to calcium sulfate. Since magnesium oxide is soluble in acids, it may not be possible to avoid forming magnesium sulfate. Gypsum and epsom salts are possible end products.

Sulfite Leaching: Option RWF-4

A Canadian patent (639,443) covers a process being used by the Aluminum Company of Canada to recover magnesium from dolomitic ores. It features the use of carbon dioxide as well as sulfur dioxide to achieve a 92% recovery of magnesia, while holding the solution of calcium to less than 3% in the magnesium solution. Sulfur dioxide from the regenerator, fortified with carbon dioxide, might be used to produce magnesium bisulfite, usable in the paper industry. Alternatively, the bisulfite could be calcined to recycle the sulfur dioxide and produce pure magnesia. Here, sulfur recovery facilities would also be required. A nonregenerative version is not included because a ready source of sulfur dioxide is lacking.

EVALUATION OF SPENT STONE DISPOSAL SYSTEMS

Solids from fluidized bed combustion plants will vary, depending on how the system is operated. Spent solids include fuel ash and dry, granular, spent sulfur sorbent. Spent sorbent composition with limestone or dolomite may consist of $CaSO_4$ and $CaCO_3$; $CaSO_4$ and CaO; $MgO \cdot CaSO_4$ and $MgO \cdot CaCO_3$; or $MgO \cdot CaSO_4$ and $MgO \cdot CaO$.

The sorbent material may be disposed of in its partially utilized form in a once-through sorbent system or regenerated for reuse in the fluidized bed combustor. Available pilot-plant test data show that calcium to sulfur (Ca/S) ratios of approximately two or greater are required to achieve 90% sulfur removal for a once-through dolomite system. It has been indicated the stone requirement for a once-through system can be significantly reduced—1.2 Ca/S ratio for 90% sulfur removal. Regeneration of the spent stone has the potential to reduce stone requirements further and, if the stone (or alternate sorbent) can be reconstituted, virtually to eliminate the need for makeup sorbent.

Direct disposal of the spent sorbent is attractive for first-generation plants. Preliminary tests indicate the disposition of the spent sorbent from fluidized bed combustors will not cause water pollution problems. Preliminary activity tests also indicate the temperature increase of the spent stone is negligible when subjected to the environment. The chemical composition of the spent sorbent from once-through and regenerative processes and its likely environmental impact require comprehensive leaching tests and activity tests to determine the chemical fate of the constituent compounds (calcium, magnesium, sulfate ion, and so on) and trace elements which may occur in the raw sorbent or which may accumulate during the combustion of the coal.

Several potential applications for the processed spent sorbent are identified as soil stabilization, landfill, concrete, refractory brick, gypsum, municipal waste treatment, and acid mine drainage.

Environmental problems associated with disposal of the spent sorbents from fluidized bed combustion systems differ favorably from those associated with disposal of lime sludges, in that they are solids and they do not possess great water solubility. Data indicate the spent dolomite (or limestone) can be used as dry landfill with known civil engineering practices for controlling structural rigidity and ground water flows. Alternatives are also available for utilization of the spent stone. In addition, the advanced sulfur removal systems being developed would minimize the quantity of spent stone available and, thus, could minimize the problem.

ASSESSMENT OF ENVIRONMENTAL IMPACT

Spent stone is composed of magnesium oxide (MgO), calcium sulfate ($CaSO_4$), calcium oxide (CaO), or calcium carbonate ($CaCO_3$), when dolomite is used, and $CaSO_4$, CaO, (or $CaCO_3$) when limestone is used. The waste stone has particle size ranges of 6.4mm (¼ inches) and smaller. Factors affecting the final composition include composition of the fuel, composition of the limestone or dolomite, and operating conditions. As mentioned, spent sorbent may be utilized as road-base aggregates, concrete and block fillers, neutralizing agents for acid mine drainage, and so on; or marketed for chemical recovery–sulfur recovery or magnesium oxide extraction. The environmental impact of the spent stone when it is dumped or used as a landfill is assessed here.

Chemical Stability

Of the above compounds, calcium sulfate and calcium carbonate are most likely to be environmentally stable and suitable for direct disposal without further processing, as the abundance of naturally occurring limestone and gypsum deposits will attest. Calcium oxide will hydrate readily to form calcium hydroxide [$Ca(OH)_2$] with release of heat [64.041 J/g-mol (15,300 cal/g-mol)] on contact with water unless it is dead-burned. Recarbonation of calcium oxide will also occur with carbon dioxide in the ambient air in the presence of water or water vapor.

Magnesium oxide is virtually insoluble (0.0086 g) and does not hydrate under atmospheric pressure except in the case of commercially prepared reactive grades. When most types of dolomite quicklimes are hydrated under atmospheric conditions, all calcium oxide components readily hydrate, but very little of magnesium oxide slakes. As a result, high-calcium quicklime slakes much more readily than does dolomitic, which usually requires pressure or long retention periods for complete hydration because of its hard-burned magnesium oxide component.

Magnesium oxide occurs infrequently in nature as the mineral periclase; it is also the end product of the thermal decomposition of numerous magnesium compounds. The physical and chemical properties of magnesium oxides vary greatly with the nature of the initial material, time and temperature of calcination, and trace impurities. The increase in density which results from increasing calcination temperature is paralleled by a decrease in reactivity.

Magnesium oxide, prepared in the temperature range of 400° to 900°C (752° to 1652°F) from magnesium hydroxide or basic magnesium carbonate, is readily soluble in dilute acids and hydrates rapidly, even in cold water. Fused magnesium oxides are virtually insoluble in concentrated acids and are indifferent to water unless very finely pulverized.

The oxides prepared below 900°C (1652°F), which are easily hydrated with water, are known as caustic-burned magnesias. The unreactive magnesias, prepared at higher temperatures, are called dead-burned or sintered magnesias. The hydration rate of various magnesium oxides is determined by the active surface area and may vary from a few hours, in the case of the reactive oxides obtained at low temperatures, to months or even years for the dead-burned grades.

It is generally recognized that the dissociation temperature for calcite is 898°C (1648°F) under atmospheric pressure in a 100% carbon dioxide atmosphere.

The temperature of dolomite, however, is not nearly as explicit. Magnesium carbonate ($MgCO_3$) dissociates at a much lower temperature—402° to 480°C (756° to 896°F). Since the proportion of magnesium carbonate to calcium carbonate differs in the many species of dolomites, the dissociation temperature also varies. Differences in the crystallinity of stone also appear to add to the disparity of data. The magnesium carbonate component of dolomite decomposes at higher temperatures than do natural magnesites.

A good average value for complete dissociation of magnesium carbonate in dolomite at 760 mm pressure in a 100% carbon dioxide atmosphere is 725°C (1337°F); the calcium carbonate component of dolomite would, of course, adhere to the above higher value representing dual-stage decomposition. As a result of these differences in dissociation points, the magnesium oxide is usually hard-burned before the calcium oxide is formed.

It follows that there may be some question about the environmental stability of the magnesium oxide component of the spent sorbent mixtures calcium carbonate/calcium sulfate/magnesium oxide. When the fluid bed boiler conditions approach 955°C (1750°F) and 1013 kPa (10 atm), it is expected that the magnesium oxide component is hard-burned and, therefore, suitable for disposal without further processing.

Environmental Impact Considerations

The environmental impact of any disposed material is a function of its physical and chemical properties and of the quantity involved. Potential water pollution problems can, in many cases, be predicted by chemical properties such as solubility, the presence of toxic metals, and the pH of leachates. Disposal of the spent stone from the fluid bed combustion system may create air pollution or an odor nuisance such as hydrogen sulfide, depending on the amount of calcium sulfide present (not expected in significant quantities). Heat may be released on hydration of calcium oxide when the calcium in the spent stone exists in the calcined state if the combustion temperature is high enough to produce fully calcined dolomite.

The first consideration when looking at potential water pollution from the solid waste disposal is the volume of leachate that will be produced. This is a direct function of the amount of water reaching the landfill. There are two possible sources of this water: rainfall and naturally occurring subsurface flow through the landfill site. Subsurface flow is a natural phenomenon which can seriously interfere with safe operation of landfills in two ways. First, it is a source of additional, potentially harmful leachate. Second, it can serve as a direct means of groundwater contamination. Prevention can be effected by a thorough geological study of the site beforehand and, if needed, installation of rerouting devices for the groundwater flow. In a similar vein, coverage of the landfill area when complete will greatly reduce, if not eliminate, the amount of leachate produced.

In order to predict leachate characteristics of a landfill, it is first necessary to describe the general features of water movement and geological considerations for this disposal method. Due to the recent surge of ecological interest in sanitary landfills for solid waste disposal, there is adequate information available.

Experiments

Leaching experiments and activity tests have been performed in order to assess the potential environmental impact of the spent stone from the pressurized fluid bed combustion process and its suitability for disposal as a landfill material. As calcium sulfate was a major constituent of the waste stone from the pressurized fluid bed combustion processes used in these experiments, a naturally occurring calcium sulfate was selected to undergo similar leaching conditions for comparative purposes.

Leaching Tests: A series of leaching experiments was designed to study leachate characteristics as functions of the varying parameters and procedures to induce leachates. The experiments were designed to measure mixing time, stone load, mixing mode, sample compaction, and run-off.

The chemical characteristics (pH, specific conductance, calcium, magnesium, sulfate, sulfide, and trace metal concentrations) of leachates induced from the waste stone sample under conditions corresponding to the severest cases were compared with leachates from a natural gypsum and with water quality standards set by several government agencies.

To reiterate, results from leaching tests on the spent stone sample from the pressurized fluid bed combustion process indicated that:

> In both the leaching-time experiment and the stone-loading experiment, calcium and sulfate dissolution plateaued at concentrations limited by the calcium sulfate solubility.
>
> The equilibrium calcium and sulfate concentrations were high, exceeding the water quality criteria. Since calcium sulfate occurs abundantly in nature as gypsum, leachates induced from a natural gypsum offer a good reference for its calcium and sulfate concentrations. It is indicated that control gypsum leachates contained approximately the same amounts of dissolved calcium and sulfate ions as the spent sample leachates. Both agreed relatively well with the calcium sulfate solubility, and both exceeded the water quality standards, 75 mg/l for calcium and 250 mg/l for sulfate.
>
> There was negligible dissolution of magnesium ions.
>
> Insignificant amounts of heavy metal ions were found in the leachates.
>
> Spent sample leachates were alkaline, with pH = 10.6 to 12.1. However, the run-off leachates showed a gradual decrease in pH with the amount of leachates passing through.

Results from leaching tests likewise indicate that the leachates had high pH, calcium, and sulfate, and negligible magnesium ions. Results also indicated that it is unlikely that the heavy metal ions in the leachates from pressurized fluid bed combustion processes will cause water pollution. The solutions were alkaline, but the run-off tests showed a gradual decrease in pH with the amount of water passing through.

Superficially, the calcium and sulfate concentrations in the leachates might suggest the possibility of a water pollution hazard from these ions, but it must be emphasized that the conditions in these experiments were much more extreme than those that would exist in actual cases, where water percolation is minimized. The fact that equally high calcium sulfate dissolution was found in control gypsum leachates under identical conditions offers a useful comparison. Additional tests using much larger quantities of material and taking into consideration geocriteria such as topography, geology, hydrology, and soil conditions of the actual landfill site would help to determine fully the environmental impact of spent stone disposal.

Activity Test: Heat release experiments were carried out on the following samples to determine their reactivity toward water:

(1) Spent sorbent from a pressurized fluidized bed combustion process—sulfated Tymochtee dolomite, –14 mesh

(2) Tymochtee dolomite, –16+18 mesh

(3) Spent sorbent from a pressurized fluidized bed combustion process—sulfated limestone 1359, –8+80 mesh

(4) Limestone 1359, –16+18 mesh

(5) Limestone 1359, –18+35 mesh

(6) Calcined limestone 1359 at 960°C (1760°F) –18+35 mesh

In each test, 3 g of stone was added to 20 ml of deionized water in a Dewar flask which had been thermally equilibrated. Iron-constantan thermocouples were used to monitor the temperature rise in the stone/water system with an Omega cold junction compensator and a digital voltmeter readout. Table 6.3 summarizes the maximum temperature rise and time required for reaching the temperature. Less than 0.2°C (0.36°F) temperature rise was found for samples 1 to 5 but a temperature rise of 55°C (99°F) was found for sample 6.

The great contrast between the calcined limestone and the other samples indicated the validity of the experiment as well as the lack of reactivity of the spent stones with water. Although it can be safely assumed that no heat pollution will result if these particular batches of spent stone are subjected to rainfall after disposal, it should be noted that the activity and heat release properties of the spent sorbent are functions of the operating conditions of the fluid bed combustion process.

TABLE 6.3: SUMMARY OF STONE ACTIVITY TESTS

Sample	Stone/H$_2$O	ΔT_{max}, °C (°F)	t_{max} (sec)
1	3 g/20 ml	0.1 (0.18)	–
2	3 g/20 ml	0.1 (0.18)	–
3	3 g/20 ml	0.2 (0.36)	5
4	3 g/20 ml	0.1 (0.18)	–
5	3 g/20 ml	0.1 (0.18)	–
6	3 g/20 ml	55 (99)	4-20

Source: PB 246 116

Evaluation of Environmental Disposition Methods

Results indicate that disposal of spent stone into the environment would probably not cause water and heat pollution. It must be remembered, however, that
the physical and chemical properties of the spent sorbent are functions of the
operating conditions of the pressurized fluid bed process and that the physical
characteristics of the specific disposal site must be judged individually in evaluating the leaching properties of the spent stone.

To assess fully the environmental impact of spent sorbent, additional tests on
stone analyses, leaching properties, heat release properties, landfill properties,
and air emission should be considered.

REGENERATION SYSTEMS
FOR SPENT STONE

The material in this chapter was excerpted from PB 231 162. For a complete bibliography see p 263.

PROCESS CONCEPTS

The economics and performance of three regeneration processes which function to regenerate the utilized SO_2 sorbent produced in the boilers of the pressurized fluid bed combustion power plant, have been studied and projected. These processes are a one-step process operated at 10 atmospheres pressure, the same one-step process operated at 1 atmosphere pressure, and a two-step regeneration process.

A sensitivity analysis has indicated the potential of the one-step process at both high and low pressures. In addition to the consideration of these continuous regeneration process operations which follow the variable power plant load, the concept of a constant load regeneration process has been examined. In this constant load case, storage capacity is situated between the boilers and the regeneration process to allow the regeneration process to operate continuously at a fixed sulfur load, or with on-and-off control. Analysis of once-through operation in which the utilized sorbent is disposed of without further processing has also been carried out to weigh the economic advantages and disadvantages of sorbent regeneration.

One-Step High-Pressure and Low-Pressure Regeneration

The one-step dolomite/limestone regeneration process consists of a single fluidized bed reaction vessel in which utilized dolomite/limestone ($CaSO_4$) from the pressurized fluid bed boiler (coal-fired) is reacted with a H_2/CO reducing gas to produce CaO and SO_2. The reaction

$$(1) \qquad CaSO_4 + \begin{pmatrix} H_2 \\ CO \end{pmatrix} \rightleftharpoons CaO + \begin{pmatrix} H_2O \\ CO_2 \end{pmatrix} + SO_2$$

requires temperatures on the order of 2000°F to produce high levels (1 to 15%) of SO_2 at reasonable rates. The SO_2 equilibrium concentration is favored by reduced pressure, being inversely proportional to the total pressure. The competing reaction

$$(2) \qquad CaSO_4 + 4\binom{H_2}{CO} \rightleftharpoons CaS + 4\binom{H_2O}{CO_2}$$

also occurs to produce the unwanted product CaS.

The regenerated dolomite/limestone is returned to the fluid bed boiler with the required amount of fresh dolomite/limestone to supplement the regenerated sorbent's reduced activity. The SO_2 stream from the regenerator is sent to a sulfur recovery plant in which sulfuric acid of elemental sulfur is produced.

The one-step regeneration process is broken down into five interrelated elements: a regenerator element, a reducing-gas producer element, a sulfur recovery element, a Claus plant tail gas handling element, and a sorbent circulation element. These elements and their relationship to one another are shown in Figure 7.1.

The high- and low-pressure processes are conceptually identical. The simple block flow diagram indicates the basic streams in the process. Fuel, air and/or steam is converted to a H_2/CO reducing gas in the reducing-gas producer element. This reducing gas is combined with tail gas recycled from the sulfur recovery element to make up the reducing gas provided to the regenerator. Utilized sorbent from the fluid bed boiler is transported to the regenerator element to produce a regenerated sorbent and an SO_2 stream. The regenerator product gas is processed in the sulfur recovery element, with a portion of the sulfur recovery tail gas being recycled to the regenerator and the remainder processed in the tail gas handling element. The performance and design of these five regeneration process elements are interrelated in a complex manner.

Two-Step Regeneration

The two-step sorbent regeneration requires two fluidized bed reactors in series to carry out the conversion of calcium sulfate to calcium carbonate. In the first step, calcium sulfate is reduced to calcium sulfide by the reaction

$$CaSO_4 + 4\binom{CO}{H_2} \rightleftharpoons CaS + 4\binom{CO_2}{H_2O}$$

at about 1500°F and 10 atmospheres. The second step converts calcium sulfide to calcium carbonate by the reaction

$$CaS + H_2O + CO_2 \rightleftharpoons CaCO_3 + H_2S$$

at about 1250°F and 10 atmospheres. The regenerated sorbent is recycled to the boilers while the H_2S stream is sent to a sulfur recovery plant. The two-step regeneration process provides a low temperature route to regenerated sorbent.

Figure 7.2 illustrates the elements of the two-step process. The process consists of seven elements: the calcium sulfate reducer, the H_2S generator, the reducing-gas producer, the CO_2 recovery, the sulfur recovery, the tail gas handling, and the sorbent circulation element.

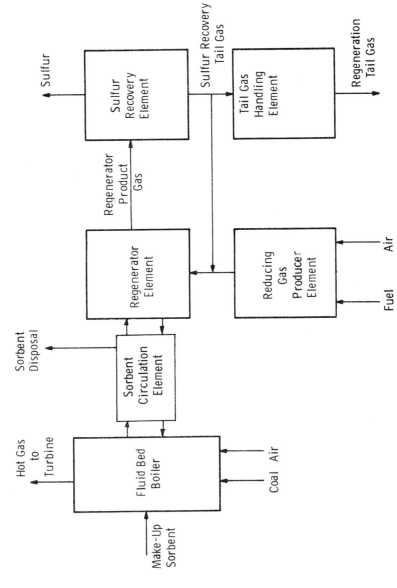

FIGURE 7.1: ONE-STEP REGENERATION PROCESS ELEMENTS

Source: PB 231 162

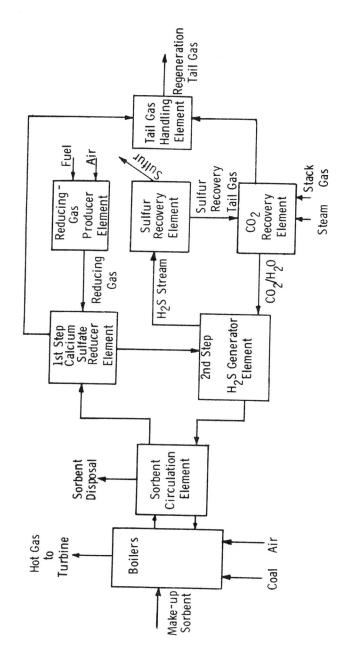

FIGURE 7.2: TWO-STEP REGENERATION PROCESS ELEMENTS

Source: PB 231 162

Reducing gas for the first-step sulfate reduction is provided by the reducing-gas producer element, while CO_2 for the second-step calcium carbonate production is provided by recovering CO_2 from stack gas and recycled sulfur recovery tail gas. Tail gases from the calcium sulfate reducer element and the CO_2 recovery element are incinerated or recycled to the fluid bed boilers in the tail gas handling element. The sorbent circulation element carries out the function of circulating the sorbent between the fluid bed boilers and the regeneration process. Elemental sulfur is recovered in the sulfur recovery element.

Constant Load

In order to eliminate the problems involved with designing the regeneration process to follow the power plant load, storage of the utilized sorbent and the regenerated sorbent between the fluid bed boilers and the regeneration process may be used to allow the regenerating process to be operated at a constant load. This concept is shown in Figure 7.3. Solid sorbent heat exchangers would be required to insure operability and to avoid large thermal efficiency losses.

The concept could be applied to either the one-step or two-step regeneration. If the power plant load fluctuates over a short time period (one week) then sufficient storage capacity can be supplied so that the regeneration process can be designed at a capacity equal to the average plant load rather than the maximum plant load required in the variable load concept. Such capability can lead to a large cost reduction in the cost of the regeneration process. Fluctuations in the power plant load may occur to the extent that the regeneration process must be designed for maximum plant capacity.

Once-Through Operation

Once-through operation with pressurized fluid bed combustion is a simplified operation because the calcium sulfate produced in the boiler is in a stable form suitable for disposal without further processing. The cost of the process is minimized by operating with maximum sorbent utilization. Problems of control are minimized by going to the once-through operation.

Wellman-Lord Stack Gas Cleaning Process

The pressurized fluid bed combustion power plant can also be operated without including a sulfur sorbent in the fluid bed boiler and instead using a final cleaning of the plant stack gas with the Wellman-Lord process. The Wellman-Lord process is a regenerative process producing elemental sulfur and has been successfully operated on a large scale. Sorbent regeneration and disposal would be eliminated by this concept at the cost of the problems associated with stack gas cleaning. Economic and environmental projections using this concept are summarized later.

Alternative Concepts

The desulfurization concepts considered for pressurized fluid bed combustion here are not meant to represent an exhaustive collection of alternatives. Many other alternatives exist. Sulfur sorbents other than limestone or dolomite might improve the process performance, economics, and environmental impact. The possibility of alternative regeneration processes for dolomite or limestone also

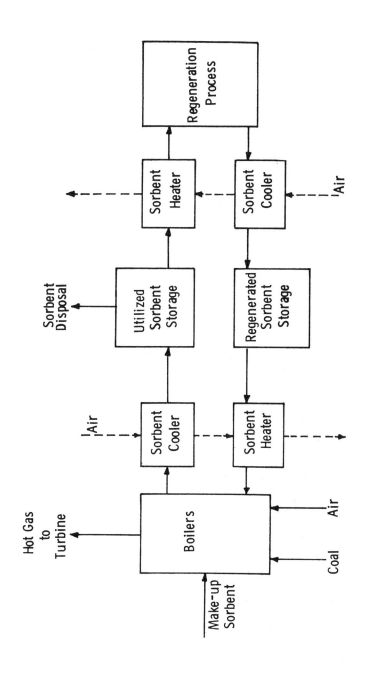

FIGURE 7.3: CONSTANT LOAD CONCEPT

Source: PB 231 162

exists. These would include both high-temperature and low-temperature concepts. Scrubbing systems other than the Wellman-Lord process could be considered to operate with the pressurized fluid bed power plant. The scrubbing process can also be operated at pressure to desulfurize the cooled gas before it enters the gas turbine. Also, other plant cycles might be considered. It would be beneficial that these alternatives, and others, be dealt with in some detail to assess their potential value.

PROCESS OPTIONS AND VARIABLES

The design options and process variables which are of importance to the proposed concepts in Figures 7.1 and 7.2 are listed below.

[1]   Reducing-gas producer fuel and process
[2]   Source and process for CO_2 production
[3]   Recovery of elemental sulfur or sulfuric acid
[4]   Recycle of process tail gas to boilers or incineration
[5]   Limestone or dolomite as sorbent
[6]   Disposal of spent sorbent before or after regeneration
[7]   Number of regenerator modules

Reducing-gas for the reducing stages of the one-step or two-step regeneration processes can be produced by gasification; partial oxidation; or reforming of coal, char, residual oil, distillate fuel oils, or methane. The source of CO_2 for the second stage of the two-step regeneration process is also a process consideration having many options. The recovery of elemental sulfur or sulfuric acid from the SO_2 or H_2S streams produced in the processes are additional options.

The tail gas from the regeneration process must be either incinerated and released to the atmosphere or recycled to the fluid bed boilers. Incineration might be followed by energy recovery (turbine expander or waste heat boiler) and final cleanup of SO_2. Both limestone and dolomite are candidates for sulfur sorbents, each one having possible advantages and disadvantages. The spent sorbent could be disposed of prior to or following the regeneration step. The choice of disposal location (before or after regeneration) can greatly affect the cost and performance of the regeneration process. The number of regenerator modules in the plant depends on the plant turndown requirements and the capabilities of the various elements to operate at reduced load.

The items in the following list represent a simplification of the complete set of variables which describe the regeneration process. Those variables having the greatest effect on the process design and performance are included.

[1]   Regenerator vessel temperature
[2]   Regenerator vessel fluidization velocity
[3]   Mol fraction SO_2 or H_2S from regenerator
[4]   Reducing-gas composition to regenerator
[5]   Weight fraction sulfur in the boiler coal, W_s
[6]   Sorbent make-up rate, m, mols Ca/mol S
[7]   Boiler sulfur removal efficiency, ϵ
[8]   Sorbent utilization in the boilers, x_s^B, and after
        regeneration, x_s^R
[9]   Sorbent purchase plus disposal cost, $/ton

[10]   Boiler conditions: temperature, pressure, excess air, calcination of the sorbent or carbonization

[11]   Process sulfur load: pounds of sulfur processed in the regeneration process per pound of coal fed to the boilers

[12]   Process sulfur load = $W_S (\epsilon - mx_S^B)$ for sorbent disposal before regeneration

[13]   Process sulfur load = $W_S (\epsilon - mx_S^R)$ for sorbent disposal after regeneration

BASE CASE DESIGNS

Parametric studies of the economics of one-step dolomite regeneration have considered a wide range of operating and design variables. In order to gain clarity and obtain the most probable regeneration process configuration and cost, base designs of the alternative regenerating processes described for use with pressurized fluid bed combustion have been prepared on the basis of the best existing experimental information.

Design Specifications

The design specifications are shown below.

[1]   Fixed coal rate to boilers—430,000 lb/hr (~635 MW)

[2]   Sorbent type—dolomite

[3]   Boiler sulfur removal—95%

[4]   Variable process sulfur load—0.01 to 0.06

[5]   Recycle of process tail gas to boilers or incineration

[6]   Recovery of elemental sulfur

[7]   Reducing-gas production from coal only (methane for sulfur recovery)

[8]   Four regenerator modules (one for each boiler module)

The designs were carried out for a fixed coal rate to the boilers rather than a fixed plant MW capacity. Dolomite is considered rather than limestone. High levels of sulfur removal are specified (though they may not be required to meet SO_2 emission regulations with all coals). Variable process sulfur loads are considered as previously listed.

The options of recycling the process tail gas to the boilers or incinerating the tail gas are considered. Recovery of elemental sulfur is specified, and coal is assumed to be the only source of reducing gas (except for the methane needed for sulfur recovery or incineration). CO_2 is recovered from stack gas and sulfur recovery tail gas in the two-step process.

Design assumptions are listed in Table 7.1. For the one-step regeneration process, the SO_2 mol fractions and the reducing gas requirements are projected from gathered data. Two SO_2 mol fractions are considered in the high-pressure case. A temperature of 1900°F is used in the low-pressure regenerator based on the difficulty of temperature control if the temperature were set at 2000°F. Boiler conditions have been assumed to be calcination conditions with the one-step regeneration process and carbonation conditions with the two-step regeneration process. Either condition may exist in the boiler depending on the boiler pressure, temperature, and excess air. The boiler conditions chosen for the two regeneration processes are the optimum boiler conditions for the respective process. If the one-step regeneration process must handle carbonated dolomite or if the two-step process must handle calcined dolomite, then the complexity and cost of the process will increase drastically.

TABLE 7.1: DESIGN ASSUMPTIONS

Process	Case	
	10 Atm	1 Atm
One-Step Dolomite Regeneration		
Regenerator pressure, atm	10	1
Regenerator temperature, °F	2000	1900
SO_2 mole fraction, %	1 and 2	10
$CO + H_2$ mole fraction in reducing gas, %	1.5 and 3	18
Sulfur recovery efficiency, %	95	95
Dolomite utilization in boiler, %	30	30
Dolomite utilization after regeneration, %	10	10
Dolomite make-up rate, moles Ca/mole S	1	1
Fluidization velocity, ft/sec	5	5
Boiler conditions	calcination	calcination
Dolomite temperature to regenerator °F	1200	1200
Calcium sulfide in regenerated sorbent, %	0	0
Two-Step Regeneration		
$CaSO_4$ reducer pressure, atm	9	
$CaSO_4$ reducer temperature °F	1500	
H_2S generator pressure, atm	11	
H_2S generator temperature, °F	1250	
H_2S mole fraction, %	3	
$CO + H_2$ mole fraction in reducing gas, %	30	
Sulfur recovery efficiency, %	95	
Dolomite utilization in boiler, %	30	
$CaSO_4$ reduced in reducer vessel, %	100	
Dolomite utilization after H_2S generator, %	10	
Dolomite make-up rate, mole Ca/mole S	1	
Fluidization velocity, ft/sec	5	
Boiler conditions	carbonation	
Dolomite temperature to $CaSO_4$ reducer, °F	1200	

Source: PB 231 162

ECONOMIC RESULTS FOR BASE CASE DESIGNS

The basis on which energy costs were computed is presented below.

[1] Mid-1973 costs
[2] Capital charges at 15%/yr
[3] O&M at 5%/yr
[4] 70% capacity factor
[5] No credit for recovered sulfur
[6] Coal at 45¢/MM Btu
[7] Methane at 80¢/MM Btu
[8] Operating water at 8¢/M gal

Capital and energy costs for the four variable load base designs were considered and the effect of credit for recovered sulfur is indicated. The costs of the constant load concept applied to each of the regeneration processes are projected. The assumption of the state of the sorbent in the boiler (calcined or carbonated) is shown to be a critical factor in the cost and performance of a given regeneration process. Finally, the plant energy costs with the regeneration processes are compared to the plant energy cost with once-through operation, as a function of the sorbent cost.

Process Capital Investment

Results of capital cost estimates are shown in Table 7.2. The table gives the capital investment in \$/kW as a function of the process sulfur load for the three regeneration processes. The two-step regeneration process requires the greatest investment of the processes considered. The sensitivity of the investment to the SO_2 mol fraction in the regenerator product is indicated by the results for the one-step process at 10 atmospheres. The investments are comparable to investments for a limestone wet-scrubber system.

TABLE 7.2: CAPITAL INVESTMENT FOR REGENERATION PROCESSES BASE DESIGNS (\$/kW)[a]

Process Sulfur Load	One-Step Process-1 Atm 10% SO_2[b]	One-Step Process-10 Atm 1% SO_2[c]	One-Step Process-10 Atm 2% SO_2[c]	Two-Step Process[c]
0.01	13.4	18.1	11.7	21.0
0.0	27.9	36.8	24.6	46.0
0.06	47.6	60.6	40.9	76.7

[a] Does not include interest during construction and fee - mid-1973 costs - 635 MW (e) plant capacity.
[b] Incineration of process tail gas.
[c] Recycle of tail gas to fluid bed boilers.

Source: PB 231 162

Plant Energy Costs

Energy costs for the total power plant are given in Table 7.3 as a function of the process sulfur load. The table indicates that the energy cost of the one-step processes at 1 atmosphere and 10 atmospheres (2% SO_2) are about the same and the energy costs of the one-step process at 10 atmospheres (1% SO_2) and the two-step process are about the same. Again, the sensitivity of the energy cost to the SO_2 mol fraction is indicated by the 1 and 2% SO_2 costs for the 10 atmospheres one-step process.

The \$2/ton dolomite cost assumed in the table may not be a realistic cost, but is not of importance when comparing regeneration processes having identical makeup rates. The figures in Table 7.3 do not include a credit for recovered sulfur. Table 7.4 indicates what this credit would amount to as a function of the process sulfur load and the sulfur cost.

TABLE 7.3: ENERGY COST FOR POWER PLANT—BASE DESIGNS (MILLS/kWh)[a]

Process Sulfur Load	One-Step-1 Atm 10% SO_2	One-Step-10 Atm 1% SO_2	One-Step-10 Atm 2% SO_2	Two-Step
0.01	11.9	12.3	11.9	12.4
0.03	12.7	13.5	12.8	13.5
0.06	13.8	15.1	14.0	15.2

[a]Based on:
- Dolomite purchase and disposal cost of $2/ton
- Capital charges of 15%/yr
- O&M of 5%/yr
- 70% capacity factor
- No credit for recovered sulfur
- Coal at 45¢/MM Btu
- Methane at 80¢/MM Btu

TABLE 7.4: CREDIT FROM SULFUR SALES (MILLS/kWh)[a]

Process Sulfur Load	Credit From Sulfur Sales, mills/kWh Sulfur at $10/ton	$20/ton	$30/ton
0.01	0.03	0.06	0.10
0.03	0.10	0.19	0.29
0.06	0.19	0.39	0.58

[a]Based on 95% sulfur recovery efficiency and 635 MM(e) plant capacity.

Source: PB 231 162

Constant Load Concept Economics

The capital investments for the constant load concept applied to each of the regeneration processes is shown in Table 7.5. It has been assumed that the fluctuations in power plant load occur over a short time period and are reproduced throughout the year so that the design capacity of the regeneration process can be proportional to the plant capacity factor. A capacity factor of 70% has been assumed.

TABLE 7.5: CAPITAL INVESTMENT FOR CONSTANT LOAD CONCEPT
($/kW)[a]

Process Sulfur Load	Investment. $/kW 1			
	One-Step-1 Atm 10% SO$_2$ [b]	One-Step-10 Atm		Two-Step [c]
		1% SO$_2$ [c]	2% SO$_2$ [c]	
0.01	15.1	22.1	17.6	24.2
0.03	29.9	45.8	37.5	52.9
0.06	48.9	73.4	59.0	86.8

[a] Does not include interest during construction and fee -- 635 MW(e) plant capacity, mid-1973 costs, 70% capacity factor.
[b] Incineration of process tail gas.
[c] Recycle of process tail gas to fluid bed boilers.

Source: PB 231 162

The energy cost corresponding to each of the constant load regeneration processes is shown in Table 7.6. The plant energy cost is substantially increased
over the variable load costs in Table 7.3 for the one-step process at 10 atmospheres and the two-step process. The energy cost of the low-pressure one-step
process is only slightly increased over the variable load concept energy cost.
The advantages to be gained from constant load operation may overweigh this
small increase in cost.

TABLE 7.6: ENERGY COST FOR CONSTANT LOAD CONCEPT
(MILLS/kWh)

Process Sulfur Load	One-Step-1 Atm 10% SO$_2$	One-Step-10 Atm		Two-Step
		1% SO$_2$	2% SO$_2$	
0.01	12.0	12.4	12.2	12.4
0.03	12.8	13.8	13.3	13.8
0.06	13.9	15.6	14.6	15.5

Source: PB 231 162

Effect of Boiler Conditions

The base regeneration process designs were evaluated with the assumption applied that the boiler conditions have provided the optimum sorbent conditions
entering the regeneration process. This means specifically calcium oxide entering the one-step regeneration process (calcination in the boilers) and calcium carbonate entering the two-step process (carbonation in the boilers). Since this
optimum behavior may not occur during actual plant operation, it is important
to note the effect of the opposite boiler conditions on the process cost and performance. Conservative process assumptions were applied to obtain the results
discussed.

Some reduction in cost may be possible by improving the process schemes. If calcination occurs in the fluid bed boilers and the two-step regeneration process is used for sorbent regeneration, then extensive carbonation will occur in the H_2S generator vessel. The exothermic nature of the carbonation reaction will disrupt the H_2S generator temperature control and the requirement for maximum H_2S concentrations in the product gas. Thus, the carbonation of the sorbent must be carried out in a separate vessel prior to the sorbent regeneration. Carbonation may be carried out by recycling plant stack gas.

Table 7.7 shows the effects of calcination boiler conditions on the energy cost of the two-step process. The results are sensitive to the rate of sorbent circulation and the extent of calcination. Table 7.7 indicates an extremely large increase in the plant energy cost compared to the results for carbonation boiler conditions in Table 7.3.

TABLE 7.7:   EFFECT OF CALCINATION BOILER CONDITIONS ON TWO-STEP REGENERATION PROCESS[a]

Process Sulfur Load	Increase in Capital,[b] $/kW	Increase in Energy Cost,[b] mills/kWh	Energy Lost,[c] % of Fuel Input	Total Energy Cost, mills/kWh
0.01	9.5	0.36	1.1	12.6
0.03	20.4	0.77	3.2	14.7
0.06	32.9	1.24	6.5	17.7

[a]Basis: $X_S^B = 0.30$, $X_S^R = 0.10$, 100% calcination of regenerated dolomite.
[b]Based on 635 MW of plant output.
[c]Total energy cost in mills/kWh is based on this efficiency loss.

Source: PB 231 162

If carbonation of the sorbent occurs in the fluid bed boilers and the one-step regeneration process is used for sorbent regeneration (high-pressure or low-pressure operation), then extensive calcination will occur in the one-step regenerator vessel. The endothermic nature of the calcination reaction will make temperature control and maintenance of high SO_2 concentration impossible. Thus, calcination must be carried out in a separate vessel prior to regeneration.

Heat for the calcination must be supplied by the combustion of coal in a separate coal combustor. The fuel required for this process step will normally be some large fraction of the fuel rate to the fluid bed boilers (depending on the sorbent circulation rate and the degree of carbonation in the boilers) so desulfurization of the hot product gas from the calcination vessel and recovery of the hot gas energy is required. Due to the large scale of this calcination process, the regeneration process would have to become an integral part of the power generation system. This would require major modifications to the fluidized bed combustion plant design concept and would greatly complicate the plant operation.

Since the fluid bed boiler conditions (temperature, pressure, excess air) change during turndown of the power plant, the sorbent may switch between calcined and carbonated states during turndown. This is an additional operating complication to be considered for regeneration processes. The boiler condition (calcination or carbonation) is a critical consideration when comparing regeneration processes. The sorbent state has a small effect upon the energy cost of once-through operation.

Process Cost Comparison

The energy costs of the four base regeneration process designs and once-through operation are compared in Figures 7.4 and 7.5 as a function of the dolomite purchase plus disposal cost. The figures consider 95% sulfur removal of a 4 wt % sulfur coal. A dolomite utilization of 30% in the boiler and a regenerative makeup rate of 1 mol Ca/mol S are considered. Once-through makeup rates of 2 and 3 mols Ca/mol S are shown. Figure 7.4 indicates the process costs for the case of sorbent disposal from the process before the sorbent is regenerated, while Figure 7.5 represents the case of sorbent disposal after the sorbent has been regenerated. The energy cost basis listed previously was applied to the base conditions also listed there and in Table 7.2 to obtain the results.

FIGURE 7.4: PLANT ENERGY COST (DISPOSAL BEFORE REGENERATION)

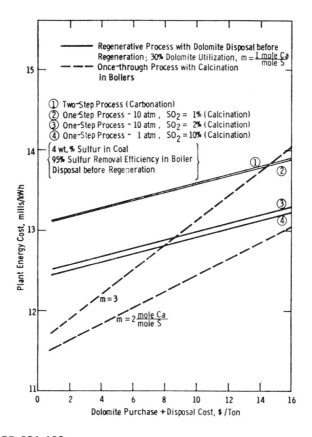

Source: PB 231 162

FIGURE 7.5: PLANT ENERGY COST (DISPOSAL AFTER REGENERATION)

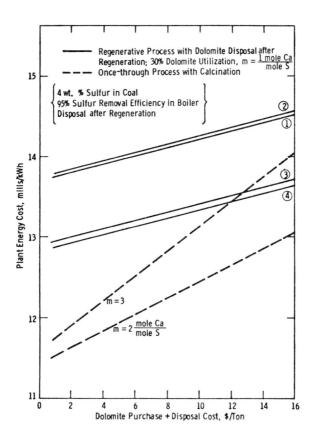

Source: PB 231 162

The case requiring sorbent disposal after regeneration yields higher energy costs than the case requiring disposal before regeneration. Disposal before regeneration gives smaller process sulfur loads than disposal following regeneration. The extent of this effect on the plant energy cost depends on the sorbent utilization before and after regeneration.

The chemical nature of the disposed sorbent wanted is a further consideration. The figures indicate a large cost advantage for once-through operation over the regeneration processes considered even for high dolomite costs. Once-through makeup rates less than 2 mols Ca/mol S may be feasible, leading to greater cost advantages. The effect of boiler conditions must also be considered. The plant operation is also simplified with once-through operation.

PROCESS PERFORMANCE

The fluidized bed plant heat rate, regeneration process temperature control, regeneration process turndown, and the overall environmental comparison of the regeneration processes are discussed here. These performance factors are necessary items to examine in order to understand the value of the various processes considered.

Plant Heat Rate

The plant heat rate is shown in Figure 7.6 for the four base design cases and for once-through operation. The heat rate is expressed as a function of the process sulfur load, $W_s (\epsilon - mx_s)$, for the regenerative processes or the factor $W_s m \times 10^{-1}$ for the once-through operation. For the once-through operation the resulting heat rate for the cases of calcination in the boiler and carbonation in the boiler are shown. Values of the regenerative process sulfur load may approach 0.06 for very high sulfur coals, while the factor $W_s m \times 10^{-1}$ for once-through operation would probably never exceed 0.025 even with very high sulfur coals. Once-through operation or the 1 atmosphere one-step process yield the lowest plant heat rates. The 10 atmosphere one-step regeneration with 1% SO_2 in the regenerator product yields the highest plant heat rate.

FIGURE 7.6: PLANT HEAT RATE WITH REGENERATION ALTERNATIVES

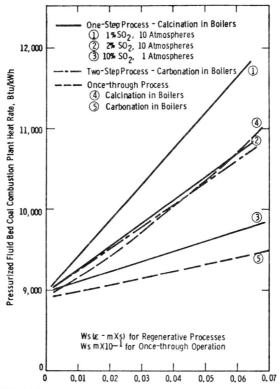

Temperature Control

The feasibility of temperature control of the regeneration processes has been investigated. The major consideration of temperature control feasibility is whether
or not the required amount of energy can be supplied or withdrawn from the regenerator vessels while maintaining the specified temperature and concentration
of SO_2 or H_2S in the regenerator product. It has already been noted that temperature control cannot be obtained in the regeneration processes if the sorbent
is calcined in the boiler with the one-step process. This problem has been handled by using separate vessels to process the sorbent before regeneration.

The two-step process is a low-temperature process compared to the one-step regeneration process, and temperature control is easily obtained in this process.
With the one-step process, the temperature of the reducing gas entering the regenerator vessel becomes higher as the SO_2 mol fraction in the regenerator increases. For the 10-atmosphere case where SO_2 mol fractions are 1 to 2%, the
temperature control is feasible without excessively high temperatures; but with
the 1-atmosphere case, where mol fractions up to 20% could be expected at
2000°F, temperature control is not possible. For this reason, the one-step process at 1-atmosphere pressure requires a reduced vessel temperature of 1900°F
and preheating of the incoming dolomite to 1900°F in a separate vessel, yielding an SO_2 mol fraction of about 10%. The highest SO_2 mol fraction which
would give feasible temperature control would be about 12 to 13%, without
having heat transfer surface in the reactors.

Process Turndown

The base designs have assumed that four regenerator vessels are used by the regeneration processes (one for each boiler module) to minimize the turndown required for each vessel. Though a variety of turndown concepts are possible, a
most attractive concept is for the regenerator temperature and pressure to be
maintained constant during turndown to yield a constant SO_2 or H_2S concentration to the sulfur recovery element during turndown.

The sorbent circulation rate would be maintained constant during turndown to
provide simpler control of the sorbent circulation equipment. The fluidization
velocity in each vessel would be proportional to the fractional load on each vessel during turndown. With four regenerator modules, no more than 50% turndown of each vessel would be required.

Turndown of other elements in the regeneration process have not been considered in detail, though the reducing-gas producer element and the sulfur recovery
element may be problem areas, and multiple modules may be required to obtain
high performance during turndown. The option of operating once-through during turndown at levels where the regeneration process performance is low should
also be considered.

Process Environmental Comparison

The four base designs for regeneration are compared with once-through operation of the pressurized fluid bed combustion power plant on the grounds of environmental factors in Table 7.8.

TABLE 7.8: COMPARISON OF PROCESSES—ENVIRONMENTAL IMPACTS[1,2]

| | One-Step 10 atm | | | | One-Step 1 atm | | Two-Step | | Once-Through | |
| | 1% SO$_2$ | | 2% SO$_2$ | | 10% SO$_2$ | | | | m = 3.0 | m = 1.5 |
	A	B	A	B	A	B	A	B	A	B
Plant heat rate (Btu/kWh)[3]	10,300	10,720	9,830	10,100	9,350	9,470	9,780	10,000	9,250	9,080
Plant energy cost (mills/kWh)[4]	13.88	14.61	13.21	13.67	13.13	13.59	13.94	14.67	13.45	12.30
Raw materials[4]										
Total coal input (tons/hr)	244	252	233	238	223	226	234	240	215	215
Dolomite input (tons/day)	1,040	1,040	1,040	1,040	1,040	1,040	1,040	1,040	3,120	1,560
Methane input (MM scf/day)	2.82	3.67	2.67	3.45	2.46	3.20	0	0	0	0
Plant exports										
Waste dolomite (tons/day)	890	770	890	770	890	770	1,080	1,070	2,700	1,640
Ash output (tons/day)[5]	585	605	560	571	535	543	562	576	516	516
Sulfur output (tons/day)	158	164	151	155	137	138	153	156	0	0
Sulfur removal efficiency (%)[6]	90	90	90	90	85	85	90	90	94	94
Percent increase in U.S. coal production relative to once-through operation for total conversion to fluid bed combustion	13.3	17.3	8.1	10.5	3.8	4.9	90	11.7	0	0
Percent of U.S. limestone/dolomite production required for complete conversion to fluid bed combustion[7]	10	10	10	10	10	10	10	10	30	15

[1] Column A for disposal before regeneration, Column B for disposal after regeneration.
[2] Basis: 430,000 lb/hr coal feed to boilers.
  Weight percent sulfur in coal = 4.6.
  Boiler sulfur removal efficiency = 95%.
  Dolomite utilization in boiler = 30%.
  Dolomite utilization after regeneration = 10%.
  Regenerative makeup rate = 1 mol Ca/mol S.
[3] Includes methane input.
[4] Other materials for regenerative processes include process water, chemicals and catalysts.
[5] 10% ash in coal.
[6] Sulfur removal efficiency is lower for the one-step process at 1 atm pressure because the process tail gas is incinerated rather than recycled to the boiler.
[7] Limestone/dolomite production taken from J.J. O'Donnell, A.G. Sliger, "Availability of Limestones and Dolomites," presented at the Second International Lime/Limestone Wet Scrubbing Symposium, New Orleans, Louisiana, November 8 to 12, 1971.

Dolomite cost = $10/ton.
Coal at 45¢/MM Btu.
Methane at 80¢/MM Btu.
Sulfur recovery efficiency = 90%.
Base design assumptions, restrictions, specifications, etc.
Once-through case assumes calcination in boilers.

Source: PB 231 162

For each base design two columns are indicated: column A for disposal of dolo-mite before regeneration and column B for disposal following regeneration. For once-through operation two fresh dolomite feed rates are considered: m = 3.0 and m = 1.5 mols Ca/mol S fed which corresponds to 31.6 and 63.3% dolomite utilization in the boilers, respectively. The design basis is summarized on the table.

The first two factors, plant heat rate and plant energy cost, indicate the overall impact of the processes on the energy picture. The plant heat rate and the plant energy cost should be lower for once-through operation than for the regenera-tive processes, even with a dolomite cost of $10/ton.

Other factors of importance are the plant raw materials and the plant outputs. Coal and methane requirements are greater with the regenerative processes than they are with the once-through operation at the cost of the increased dolomite consumption required for once-through operation. Ash output is comparable to dolomite waste for the regenerative processes. Sulfur removal efficiency is ex-pected to be greater for once-through operation than for regenerative operations, as indicated in the table, because of reduced chemical processing.

If the total United States electrical utility industry were to convert completely to pressurized fluid bed combustion, then the rate of coal consumption would be greater for conversion to regenerative operation than for conversion to once-through operation. The relative rate of coal consumption for regenerative and once-through operations is shown in Table 7.8 and ranges from 4 to 18% greater for regeneration.

Consumption of methane is another factor to be considered. The price paid for this reduction in coal and methane consumption (in terms of environmental im-pact) is indicated as the last factor in the table. For complete conversion of the United States electric utility industry to pressurized fluid bed combustion, about 10% of the present United States limestone/dolomite production would be required for regenerative operation and 15 to 30% for once-through operation.

ASSESSMENT OF REGENERATION SYSTEMS

Economic Factors

The plant energy cost for pressurized fluid bed combustion, utilizing dolomite regeneration and once-through operation, is compared to the plant energy cost for a conventional coal-fired power plant using limestone wet scrubbing for pollu-tion control in Figure 7.7. Curves for the one-step process with 1% SO_2 (10 at-mospheres) and 10% SO_2 (1 atmosphere) are shown.

The costs for the one-step process with 2% SO_2 and the two-step process are not shown because they have costs nearly identical to the costs for the one-step with 10% SO_2 (1 atmosphere) and the one-step with 1% SO_2 (10 atmospheres), re-spectively. The cases of dolomite disposal before and after the regeneration are shown. Both limestone and dolomite costs have been taken as $2/ton. The ef-fect of dolomite cost is shown in Figures 7.4 and 7.5.

FIGURE 7.7:   COMPARISON OF FLUID BED POWER PLANT WITH CON-
VENTIONAL POWER PLANT

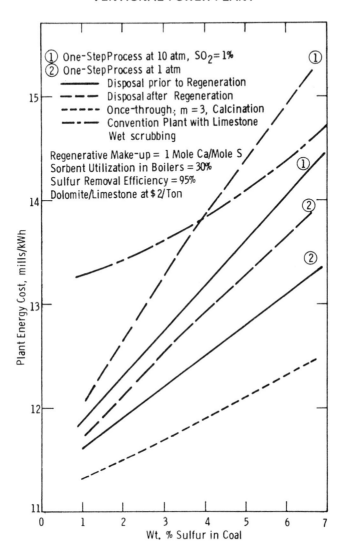

Source: PB 231 162

Figure 7.7 indicates that the energy cost of the conventional plant increases at
a slower rate with increased coal sulfur content than does the cost of the fluid
bed combustion plant. The cost of the conventional plant is greater than the
fluid bed combustion plant for all cases except the one-step process with 1% SO_2
with dolomite disposal following regeneration. The costs for the one-step process with 1% SO_2 with disposal before regeneration and the two-step process are
comparable to the conventional plant energy cost at high-sulfur contents.

The energy cost reduction with the one-step process with 10% SO$_2$ (1 atmosphere) and the one-step process with 2% SO$_2$ is indicated to be about 1 mill/kWh or greater. The cost reduction with once-through operation is shown to be about 2 mills/kWh with a low dolomite cost of $2/ton. The once-through cost reduction compared to the conventional plant will be about 1 mill/kWh with dolomite and limestone costs of $10/ton.

A general comparison of energy costs between fluidized bed combustion with dolomite regeneration, fluidized bed combustion with once-through operation, fluidized bed combustion with Wellman-Lord stack gas cleaning, a conventional power plant with limestone wet scrubbing, and a conventional power plant with Wellman-Lord stack gas cleaning is shown in Figure 7.8.

FIGURE 7.8: GENERAL COMPARISON OF ENERGY COSTS

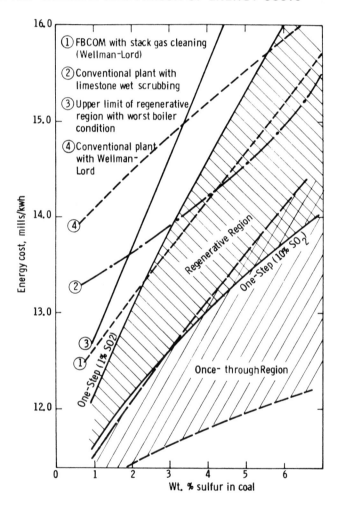

Source: PB 231 162

Table 7.9 indicates the conditions applied in Figure 7.8, with costs ranging between optimistic conditions and pessimistic conditions for regenerative and once-through fluid bed combustion power plants.

TABLE 7.9: GENERAL COMPARISON

- FBCOM with Regeneration

    Optimistic Case   Pessimistic Case

Optimistic Case	Pessimistic Case
Optimum boiler condition	Optimum boiler condition
Ca/S molar ratio, m = 1/2	Ca/S molar ratio, m = 1.5
Sorbent at $10/ton	Sorbent at $10/ton
Disposal before regeneration	Disposal after regeneration
No sulfur credit	No sulfur credit

- Once-Through

Optimistic Case	Pessimistic Case
Ca/S molar ratio, m = 1.2	Ca/S molar ratio, m = 3.0
Carbonation in boiler	Calcination in boiler
Sorbent at $10/ton	Sorbent at $10/ton

- Conventional Power Plant with Limestone Wet Scrubbing

- FBCOM Plant with Stack Gas Cleaning (Wellman-Lord)

- Conventional Power Plant with Wellman-Lord Stack Cleaning

Source: PB 231 162

Figure 7.9 indicates the effect of the plant capacity factor on the energy cost of regenerative and once-through fluid bed power plants, while Figure 7.10 shows the break-even cost of clean fuel as a function of the plant capacity factor. Evidently, the once-through operation becomes more economically attractive than the regenerative operation as the plant capacity factor decreases.

The pressurized fluid bed combustion plant using the Wellman-Lord stack gas cleaning process has many potential improvements in efficiency, because no sorbent is present in the fluid bed boiler, which may reduce the energy cost below that indicated in Figure 7.10. The cost should be attractive compared to a conventional plant.

In general, the potential exists for pressurized fluid bed combustion power plants to reduce the cost of electrical energy generated by conventional power plants by 5 to 15% if the regeneration and once-through performance projected can be realized in commercial application.

FIGURE:7.10: FBCOM WITH LOW-SULFUR COAL

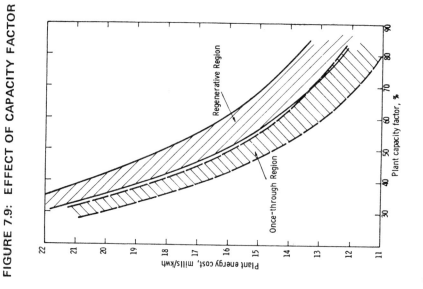

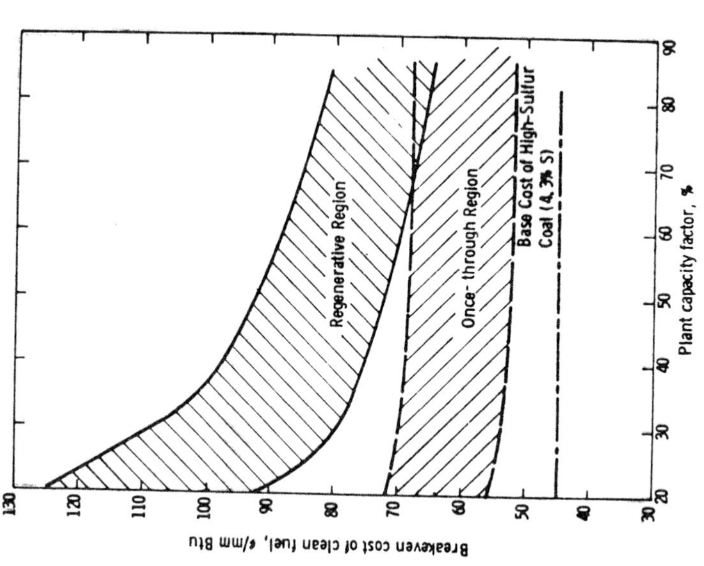

Source: PB 231 162

FIGURE 7.9: EFFECT OF CAPACITY FACTOR

Source: PB 231 162

Environmental Factors

Tables 7.10, 7.11 and 7.12 compare environmental factors for alternative power plant concepts. Resources and wastes are listed in the tables. Table 7.10 compares regenerative fluid bed combustion and once-through operation. Table 7.11 compares once-through operation of the fluid bed combustion power plant to the fluid bed combustion plant with Wellman-Lord stack gas cleaning. Table 7.12 compares the fluid bed combustion plant with Wellman-Lord stack gas cleaning to a conventional coal-fired plant with Wellman-Lord stack gas cleaning.

As shown previously, regenerative operation requires more coal and methane than once-through operation, while once-through operation requires more dolomite disposal. The net environmental effect of these factors must be considered with great care. Once-through dolomite waste levels may be reduced to values nearly as low as with regenerative operation by obtaining high utilization. $CaSO_4$ is an inert material having many potential uses.

The Wellman-Lord process requires more coal and makeup NaOH than the once-through process. The liquid disposal stream indicated in Table 7.11 for the Wellman-Lord process can be eliminated by evaporation to dryness using a low-pressure stream. NaOH manufacture requires the expenditure of large amounts of energy and is equivalent to about a 2% efficiency reduction in the fluid bed combustion power plant using the Wellman-Lord process.

The fluidized bed combustion plant using the Wellman-Lord stack gas cleaning process compares favorably with a conventional coal-fired power plant using the Wellman-Lord process. Further improvement in the performance of the fluid bed combustion plant using the Wellman-Lord process is possible.

Base Design Feasibility

While both optimistic and pessimistic cost assumptions have been made throughout the base regeneration process designs, it is assumed that all the process steps are commercially feasible. Some process steps are questionable; the questionable areas in the regeneration process are such as will lead to reduced power plant availability and high maintenance costs compared to the once-through operation. Higher energy costs than have been indicated for the process base designs may also result. Descriptions of regenerator units are illustrated in the second chapter.

REGENERATION EFFECTS OF COMBUSTION OF LOW-SULFUR COALS

Considerations discussed so far have been with the fluid bed combustion power plant operated regeneratively or once-through with coals having sulfur contents greater than 1 wt % and requiring high sulfur removal (~95%). The case of low-sulfur coals which require little or no desulfurization is also of importance since such coals are attractive fuels and are still found in some areas of the United States.

The capital costs of the pressurized fluid bed combustion plant and the conventional power plant with limestone wet scrubbing are shown in Table 7.13. A 1 wt % sulfur coal is considered with 70, 50 and 0% desulfurization required. The plant energy cost is shown in Table 7.14 for the same cases.

TABLE 7.10: COMPARISON OF REGENERATION WITH ONCE-THROUGH SYSTEM[a]

Item	One-Step 1 atm	One-Step 10 atm (2% SO_2)	Two-Step	Once-through
Resources				
Plant heat rate, (Btu/kWh)	9,350	9,830	9,780	9,080
Coal (tons/hr)	223	233	234	215
Methane (MM scf/day)	2.46	2.67	0	0
Dolomite (tons/day)	1,040	1,040	1,040	1,560
Sulfur (tons/day)	145	151	153	0
Wastes				
Air pollution				
Overall SO_2 removal, %	90	90	90	94
NO_x, lb NO_2/10^6 Btu	0.1-0.2	0.1-0.2	0.1-0.2	0.1-0.2
Particulates, lb/10^6 Btu	0.05-0.15	0.05-0.15	0.05-0.15	0.05-0.15
Thermal, relative	1	1	1	1
Solids, tons/day				
dolomite	890	890	1,080	1,640
ash	535	560	562	516

[a]Basis: 635 MW(e), 4.6% Sulfur in Coal, 95% boiler sulfur removal

Source: PB 231 162

TABLE 7.11: COMPARISON OF ONCE-THROUGH WITH STACK GAS CLEANING SYSTEM[a]

Item	Once-Through Fluid Bed Combustion Plant m=1.55	Fluid Bed Combustion Plant with Wellman-Lord
Resources		
Plant heat rate, Btu/kWh	9,080	9,700
Coal, relative	1	1.08
NaOH, tons/day	0	20
Sulfur, tons/day	0	135
Methane, MM scf/day	0	2.45
Dolomite, tons/day	1,560	0
Wastes		
Air pollution		
Overall SO_2 removal efficiency, %	94	90
NO_x, lb $NO_2/10^6$ Btu	0.1-0.2	0.1-0.2
Particulates, lb/10^6 Btu	0.05-0.15	?
Thermal, relative	1	1.08
Liquid, tons/day (ft^3/day)	0	300 (10,000) - sodium sulfate, sodium thio sulfate, polythionates solution with some particulates
Solids, tons/day (ft^3/day)	1,640 (14,000) - dolomite 516 - ash	560 - ash

[a]Basis: 635 MW(e), 4.6% sulfur in coal, 95% boiler sulfur removal.

Source: PB 231 162

TABLE 7.12: COMPARISON OF STACK GAS CLEANING ON CONVENTIONAL AND FBB PLANTS[a]

Item	Fluid Bed Combustion Plant with Wellman-Lord	Conventional Power Plant with Wellman-Lord
Resources		
Plant heat rate, Btu/kWh	9,700	10,000
Coal, relative	1	1.03
NaOH, tons/day	20	20.6
Sulfur, tons/day	135	140
Methane, MM scf/day	2.45	2.55
Wastes		
Air pollution		
Overall SO_2 removal efficiency, %	90	90
NO_x, lb $NO_2/10^6$ Btu	0.1-0.2	0.05-0.15
Particulates, lb/10^6 Btu	?	?
Thermal, relative	1	1.03
Liquid, tons/day	300	309
Solids, tons/day	560-ash	577-ash

[a]Basis: 635 MW(e), 4.6% sulfur in coal, 95% boiler sulfur removal

Source: PB 231 162

TABLE 7.13: LOW-SULFUR COAL COMBUSTION—INVESTMENT ($/kW)

% Desulfurization of 1% Sulfur Coal	Fluid Bed Combustion		Conventional Plant with Limestone Wet Scrubbing
	Once-through	Regenerative[a]	
70	265.6	278.5	331.4
50	265.0	272.5	330.5
0		260.5	316.3

[a]Based on the one-step regeneration process at 10 atmospheres with 1% SO_2.

Source: PB 231 162

TABLE 7.14: LOW-SULFUR COAL COMBUSTION—ENERGY COST (MILLS/kWh)

% Desulfurization of 1% Sulfur Coal	Fluid Bed Combustion		Conventional Plant with Limestone Wet Scrubbing
	Once-through	Regeneration[a]	
70	11.31	11.86	13.36
50	11.24	11.58	13.29
0		11.12	12.51

[a]Based on the one-step regeneration process at 10 atmospheres with 1% SO_2.

Source: PB 231 162

As was indicated in Figure 7.8, the cost differential between fluidized bed combustion and conventional power generation becomes larger as the coal sulfur content is reduced. An 11% reduction in energy cost is projected for the case of no desulfurization. With no desulfurization required, many improvements in the design of the fluidized bed combustion processes are possible.

PARTICULATE CONTROL SYSTEMS

The material in this chapter was excerpted from PB 246 116. For a complete bibliography, see page 263.

TURBINE TOLERANCE FOR PARTICULATES

Estimates of particle size distribution from fluidized bed combustors produced the distribution shown in Figure 8.1 (curve 1). In addition, two progressively finer distributions (curves 2 and 3) have been used in economic sensitivity analyses. This study considers the limitations imposed by particulates in fluidized beds.

Erosivity

As a first approximation, the turbine erosion damage due to the impact of a single particle is proportional to the kinetic energy of the particle. The erosivity of a dust particle may thus be defined by the product of its kinetic energy and its probability of impacting. To obtain a measure of the erosivity of a dust suspended in a gas stream it is necessary to total the erosivity of all the particles contained in a unit volume of gas. This erosivity can be used to estimate acceptable levels of particulates in gas-turbine expansion gas.

Turbine Specifications

Typical specifications for gaseous fuels for gas turbines require that dust loadings be kept below 0.023 g/m^3 (0.01 gr/scf); that 95% of the dust be less than 20 μm; and that no particles be greater than 100 μm. These specifications apply to high heating-value fuels and must be adjusted for expansion gas, as the important consideration in turbine operation is the particulate concentration in the expansion gas. The adjustment of specifications must account for the dilution of the fuel gas by excess air during combustion. High heating-value gas requires approximately a fifty-times dilution. Consequently, the particulate loading in the expansion gas on this basis should be restricted to one-fiftieth of that specified for

149

high heating-value gas, in other words, 0.0005 g/m³ (0.0002 gr/scf).

FIGURE 8.1: PARTICLE SIZE DISTRIBUTION ASSUMED FOR DUST ELUTRIATED FROM GASIFIER

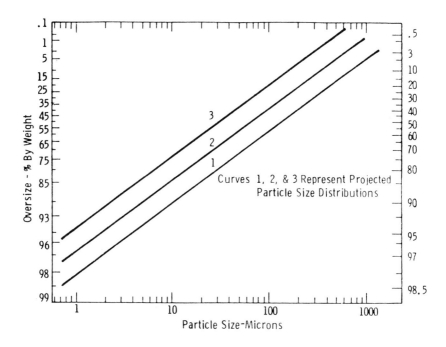

Source: PB 246 116

The erosive nature of a dust which would comply with this specification is dependent on the loading, the particle size distribution, and its velocity relative to turbine components.

It should be noted that particles below two microns are not considered erosive in this case, as their efficiency of impaction is negligible. This illustrates the turbine tolerance considerations and the requirements which must be considered for the particulate removal system.

DUST COLLECTION EQUIPMENT

Cyclones

Cyclones are able to handle heavy dust loading and high gas throughput at reasonable cost and low-pressure drop. They are not prone to plugging, but their internal surfaces are subject to erosion.

FIGURE 8.2: MULTICYCLONE ARRANGEMENTS

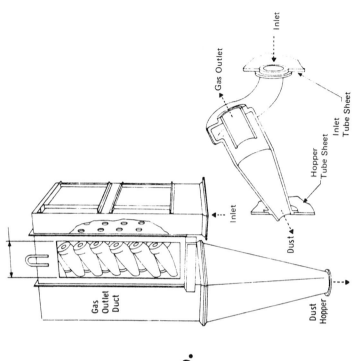

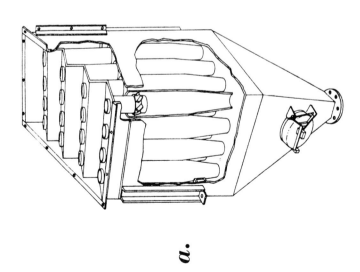

(a) Swirl vane inlet
(b) Tangential inlet

Source: PB 246 116

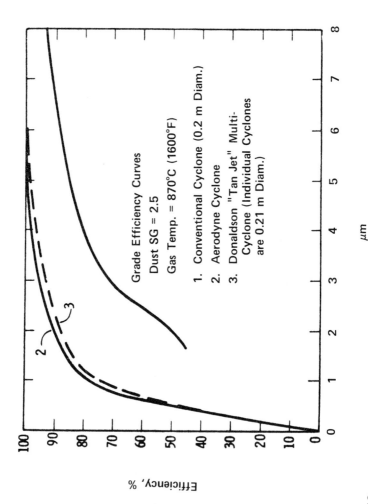

FIGURE 8.3: CYCLONE GRADE EFFICIENCY CURVES

Grade Efficiency Curves

Dust SG = 2.5

Gas Temp. = 870°C (1600°F)

1. Conventional Cyclone (0.2 m Diam.)
2. Aerodyne Cyclone
3. Donaldson "Tan Jet" Multi-Cyclone (Individual Cyclones are 0.21 m Diam.)

Cyclones of conventional design lose efficiency as the cyclone diameter is increased. Thus, for high efficiency with high gas flow, banks of small cyclones are manifolded together in a parallel flow arrangement (Figure 8.2). If the cyclones are connected directly to a common dust hopper, secondary flows between individual units reduce the operational efficiency of these units. Consequently, performance may be improved by withdrawing the dust individually from each unit cyclone.

High-efficiency cyclones which employ secondary gas flows to improve collection efficiency are also available both in large individual units and in multicyclone configurations.

Conventional Cyclones: A conventional design high-efficiency cyclone, 203.2 mm (8 in) diameter and with a capacity of 0.07 m^3/s (150 cfm) has a grade efficiency curve as shown in Figure 8.3 (at fluidized bed combustor conditions). Units of this size can be incorporated into a multicyclone design with little loss in efficiency.

Shell Collector: Shell Oil uses high-temperature multicyclones to collect particulates from cat cracker off-gases before they are expanded through gas turbines. The performance of these units is similar to that of small conventional cyclones.

Rotational Flow Cyclones: Rotational flow cyclones (e.g. Aerodyne) employ a secondary gas flow. Data for fractional efficiency show extremely high collection efficiency for particles smaller than 5 μm when a clean secondary gas flow is used (curve 2, Figure 8.3). It is difficult, however, to have such an arrangement in a fluidized bed boiler plant wherein the secondary gas flow would be provided by using a portion of the dirty gas stream.

Tests of rotational flow cyclones with a dirty secondary gas flow have shown poor collection efficiency for particles 5 μm and smaller. Unless this performance can be radically improved the units cannot be considered for use in fluidized bed gas-turbine power plants.

Donaldson Tan Jet Cyclone: The Donaldson Company has a multicyclone system which is significantly more efficient than conventional cyclones in collecting particles in the 0-to-5 μm range (Figure 8.3). This unit requires a clean secondary gas flow equivalent to 15% of the primary flow. The secondary flow should be at the same temperature as the primary flow to prevent thermal stressing of the assembly and to maintain the flow field at its design condition.

Granular Bed Filters

Three filtration systems have possibilities for fluidized bed combustors. They are: granular beds; porous metals; and porous ceramics. In the granular bed filters, beds of granules should, in principle, be able to filter dust particles from gas streams in much the same way as do beds of fibers.

Fiber filtration studies show that very high efficiencies can be achieved over a broad range of particle size. Unfortunately, fiber-bed filters cannot easily be cleaned and this limits their utility.

Granular beds are somewhat easier to clean, may be operated at high temperature, and have the potential to achieve the same high efficiencies as fiber filters. A 1970 review of the literature by Squires and Pfeffer (1) indicates that several filters have been operated with collection efficiencies better than 90% on dusts down to 2 μm particle size.

The transient behavior of granular-bed filters while collecting dispersed fly ash has been studied more recently indicating that high efficiencies are possible with clean filters, but performance deteriorates as the dust content of the filter increases.

An attempt has been made by Ducon Co. to commercialize a continuously operating granular-bed filter and its performance has been studied on cat cracker regenerator off-gas (2). The manufacture of this unit is undoubtedly too costly for it to compete with alternative collectors. Another promising technique by Squires uses a panel-bed filter, but thus far limited data are available on its performance.

Mode of Operation: Although it appears that impaction collection dominates in determining the efficiency of a granular filter, two operating regimes exist: one in which impaction predominates and particles are collected in the interstices of the filter, the second in which the initial collection at the filter surface produces a filter cake. This cake acts as a filter aid, retaining essentially all of the collected solids at the filter surface.

The first operating mode is normally associated with high gas velocities and, consequently, with high-throughput filters. When a filter operates for an extended period under these conditions, the operating efficiency will initially remain constant with time. The bed will become progressively saturated with dust, however, and as the saturation zone extends through the bed, the collection efficiency will decline.

Saturation apparently occurs when dust deposits on impaction sites profile the gas flow, avoiding sharp changes in direction and minimizing the effects of inertial deposition. It is most likely that there is also a dynamic equilibrium between deposition and reentrainment in saturated zones; thus analysis of the transient behavior of beds is complex.

The second operating mode, formation of a filtering layer, is essentially the same as is encountered in bag filters and in some porous metal filter installations (3). Although some dust leakage may be expected through the clean filter, once the filter cake is established filtration is almost absolute. As the cake builds upon the filter surface, there is a consequent increase in pressure drop. At some point, the pressure drop becomes too great; it is necessary to remove the filter cake and repeat the cycle.

Dust Removal: When dust is deposited throughout the filter, it may be removed by one of two methods:

(1) Fluidizing the bed with a reverse flow of gas. This is limited to horizontal beds and requires some secondary collector for cleaning the flushing gas.

(2) Removing both the granules and the dust from the filter and re-packing the filter with fresh granules. This is normally limited to vertical-panel filters. The package may be intermittently dumped or continuously removed in a cross flow arrangement.

When the dust accumulates in a cake on the filter surface, the cake may be removed by a reverse flow of gas. If the filter is a vertical panel, the flushing gas should be supplied as a sharp puff at high pressure. This will lift the cake from the surface and deposit it in a secondary collector. If the filter is arranged horizontally, the reverse flow of gas is used to fluidize the bed and elutriate the dust deposits.

Capacity: A 0.033 m (1.3 in) deep bed of 1,500 μm particles has a capacity of approximately 2,000 g/m^2 (3,000 gr/ft^2) before the collection efficiency begins to fall. This figure may be used as the safe operating limit for the deeper beds which would be used in a fluidized system.

Ducon Filter: Operation — The Ducon filter employs screens to retain the granular bed while permitting removal of the collected dust by blow-back techniques. The arrangement and operation of the unit are illustrated in Figure 8.4.

FIGURE 8.4: DUCON SAND BED FILTER

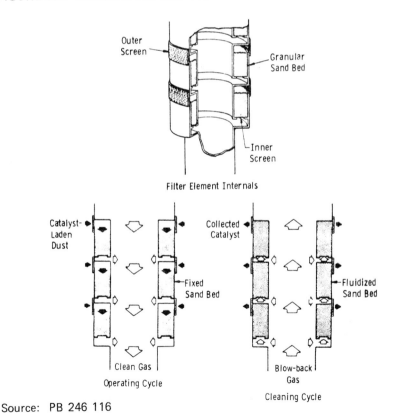

Source: PB 246 116

When being filtered, the dirty gas passes through the outer screen and down through the granular sand bed. When the bed has accumulated sufficient dust, a short blast of high-pressure gas is used to reverse the flow. The bed flexes, is fluidized, and the dust is carried from the bed.

Performance — This unit normally operates with 0.063 m (2½ in) deep bed of 760 μm sand. The usual superficial gas velocity is between 0.15 and 0.45 m/s (0.5 and 1.5 ft/s).

Performance correlations (2) show that the instantaneous efficiency of the filter improves as the quantity of dust in the bed increases. This implies that collection is enhanced by the formation of a filter cake on the bed surface. The overall efficiency, however, is considerably lower than would be expected if a coherent cake were present.

For example, when operating on cat cracker emissions, the efficiency was normally around 95%. By using a finer grade of sand, this performance can be improved. Available data on sintered metal filters (3) indicate that 100% collection down to 1 μm can be achieved with granules of 100 μm diameter and velocities around 0.03 m/s (1 ft/s). No grade efficiency curves for the filter are available.

Operating Problems —

    (1) Leakage — Dust leaks through the filter beds during the blow-back phase of the operating cycle. This markedly reduces the efficiency of the unit.

    (2) Plugging — If finer granules are used, finer retaining screens will be required. Experience has shown that fine screens have a tendency to accumulate dust deposits and become plugged.

Squires Panel-Bed Filter: Operation — The Squires panel-bed filter consists of a vertical bed of granules held in place by louvered walls which resemble Venetian blinds. The unit operates with superficial gas velocities up to 0.15 m/s (0.5 ft/s). It is indicated that collection efficiency increases with superficial velocity.

The filter is cleaned intermittently by a sharp puff-back of gas which lifts the filter cake, along with a small quantity of granules, from the surface and deposits it in a collecting hopper. Fresh granules are supplied from the top of the panel to make up for losses during the puff-back.

Performance — The efficiency of the Squires filter is not well known; however, filtration efficiencies of better than 99% have been recorded on redispersed fly ash (1), with a filter consisting of 0.025 m (1 in) sand.

Operating Problems —

    (1) Size — If the equipment operates at low gas velocities, a large filter area is needed for treating commercial quantities of high-temperature gas.

    (2) Solids handling — Distributing and collecting the filter granules produces mechanical difficulties.

Plugging — The cleanup cycle for this filter will only remove surface accumulations of dust. It is possible that with long operating schedules dust will penetrate the filter panel and block the interstices.

Cross Flow Filters: Operation — Panel bed filters may be operated with a continuous downward movement of the column of solids, clean granules being introduced at the top and a mixture of dust and granules being removed from the bottom. This is consistent with high-gas velocity/impaction collection operation.

For the fluidized bed boiler, the spent limestone granules could be used as the filter medium (Figure 8.5), providing a continuous supply of clean, hot, filter granules. As the stone is to be dumped anyway, this arrangement does not produce any secondary disposal problems.

FIGURE 8.5: CROSS FLOW FILTER ARRANGEMENT

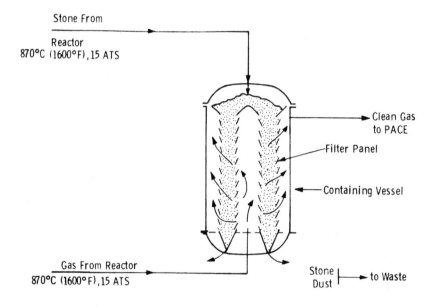

Source: PB 246 116

Performance — Dorfan (4) has built cross flow filters which use 0.013 m (½ in) to 0.038 m (1½ in) granules and claim 98% collection efficiency for 2 to 10 μm dust carried at a superficial velocity of 1.8 m/s (6 ft/s).

Zahradnik et al (5) report some preliminary results for a cross flow bed of 1,600 μm (1/46 in) alkalized alumina. They show 99% collection efficiency for redispersed fly ash carried at 0.15 m/s (0.5 ft/s).

Operating Problems —

(1) Solids flow — Solids flow in a conventional panel bed retained
by louvers will create dead zones near the filter surfaces.
These zones would eventually saturate with dust and possibly
plug.
(2) Reentrainment — Relative motion of the filter granules will
cause dust reentrainment in the gas stream. Unless the bed
is sufficiently wide in the direction of gas flow, poor performance will result.
(3) Solids handling — Solids distribution to the filter panels presents mechanical difficulties.
(4) Erosion — Solids retaining elements will be subject to erosion
by the moving granular bed.

Porous Metal Filters

Porous metal filters have been used commercially for more than 15 years to filter catalyst fines from effluent gases in fluidized bed reactors. They can be
operated with virtually 100% efficiency on particles down to one micron on a
continual, cyclic basis.

Porous metals have two serious operating problems: [1] They are limited to
operation at temperatures below 538°C (1000°F); and [2] They require a clean,
high-temperature gas supply for the cleanup portion of the operating cycle.
Porous metals are also costly: a budget estimate is $1,000 per square meter of
filter surface.

Porous Ceramic Filters

Porous ceramic filters may in principle operate in a way similar to that of porous metal filters. At high gas flows, however, vibrational strains lead to cracking
in the ceramic elements. In addition, sealing ceramic elements in a manifold is
difficult when high-temperature operation is required.

REFERENCES

(1) A.M. Squires and R. Pfeffer, "Panel Bed Filters for Simultaneous Removal of Fly Ash
and Sulfur Dioxide," *J.A.P.C.A.,* 20, 534, 1970.
(2) B. Kalen and F.A. Zenz, "Pollution Control Operations: Filtering Effluent from a
Cat-Cracker," *C.E.P.,* 69, 67, 1973.
(3) D.B. Pall, "Filtration of Fluid Catalyst Drives from Effluent Gases," *I.E.C.,* 45, 1197,
1953.
(4) R.H. Perry and C.H. Chilton, eds. *Chemical Engineering Handbook,* Fifth Edition,
New York, McGraw-Hill Book Company, 1973.
(5) R.L. Zahradnik, J. Anyigbo, R.A. Steinberg, and H.L. Toor, "Simultaneous Removal
of Fly Ash and Sulfur Dioxide from Gas Streams by Shaft Filter Sorber," *Env. Sci.
Tech.,* 4, 665, 1970.

TRACE ELEMENT STUDIES

The material in this chapter was excerpted from *Environmental Science & Technology,* PB 237 366; PB 237 754; PB 246 116 and ANL/ES–CEN–1008. For a complete bibliography see p 263.

Studies have been made of the distribution and potential emissions of biologically toxic trace elements during the fluidized bed combustion of coal. Interest in the emissions of such elements as Hg, Be, Pb, and F has increased considerably in recent years. Although these elements are present at very low levels in fossil fuels, the large consumption of these fuels in the United States represents an annual potential emission release of up to several thousand tons for some of the elements.

Since most of the trace elements in coal are likely to be found in the sorbent used for SO_2 removal, there is some interest in comparing the behavior of the trace elements in a fluidized bed combustor with conventional, coal-fired power stations. Relative to conventional boilers, the lower combustion temperatures of the fluidized bed boiler and the presence of additive for SO_2 retention may serve to retain potentially volatile trace elements in particulate form.

Combustion experiments have been performed to study the distribution of trace elements in the combustion system. The approach is to make mass balances around the combustion system for as many of the trace elements as possible within the economic limitations of the study. This involves sampling and analyzing all the solid materials charged to or recovered from the combustion system, including particulate matter entrained in the flue gas. The flue gas is also analyzed for the more volatile trace element species, Hg and F^-.

Table 9.1 summarizes concentration estimates for potential pollutants from coal-fired fluidized bed combustion. These estimates are probably good only to within an order of magnitude. The term "gas phase" includes gases, vapors, and very fine particulates ($<2\ \mu$); the term "solid phase" includes the agglomerated bed material and coarse particulates ($>2\ \mu$), which should be collected by conventional particle control devices.

TABLE 9.1: ESTIMATED CONCENTRATION RANGES OF POTENTIAL
POLLUTANTS FROM COAL-FIRED FLUIDIZED BED COMBUSTION

Gas phase

One hundred parts/million	CH_4, CO, HCl, SO_2, NO
Ten parts per million	SO_3, C_2H_4, C_2H_6
One part per million	HF, HCN, NH_3, $(CN)_2$, COS, H_2S, H_2SO_4, HNO_3, F, Na
One part per billion	Diolefins, aromatic hydro-carbons, phenols, azoarenes, As, Pb, Hg, Br, Cr, Ni, Se, Cd, U, Be
One-tenth (0.1) part per billion	Carboxylic acids, sulfonic acids, polychlorinated bi-phenyls, alkynes, cyclic hy-drocarbons, amines, pyridines, pyrroles, furans, ethers, esters, epoxides, alcohols, ozone, alde-hydes, ketones, thiophenes, mercaptans

Solids

One part per million	Al, Ca, Fe, K, Mg, Si, Ti, Cu, Zn, Ni, U, V
One part per billion	Ba, Co, Mn, Rb, Sc, Sr, Cd, Sb, Se, Ca
One-tenth (0.1) part per billion	Eu, Hf, La, Sn, Ta, Th

Source: *Environmental Science & Technology,* March 1977

USE OF SELECTED METHODS FOR ANALYZING TRACE ELEMENTS

The elements of interest, in order of priority as pollution agents as determined by the EPA, are listed in Table 9.2. Also listed are procedures believed to be the most applicable for analyzing these elements at trace concentration levels. Ideally, a method of analyzing trace elements in coal combustion products should [1] determine a large number of elements of interest simultaneously, [2] require relatively little sample preparation, [3] be capable of automation, [4] produce an output compatible with computerized data processing, and [5] be rapid.

Of the methods listed in Table 9.2, emission spectroscopy is capable of measuring with suitable sensitivity all elements, except fluorine and phosphorus. However, if this method is to meet the criteria of rapid output and computerized data processing, a rather sophisticated (and expensive) instrument equipped with direct-reading detectors would be necessary. Calibration of the instrument using materials of composition very close to those being analyzed (a time-consuming effort) also would be necessary. The resulting system would be highly specific to coal, ash, fly ash, etc., would be very rapid, and would have an accuracy of perhaps 10 to 100%, varying for different elements.

X-ray spectrometry is another method capable of simultaneously measuring most of the elements specified. It is capable of measuring all elements of atomic

TABLE 9.2: APPLICABILITY OF SELECTED ANALYTICAL METHODS FOR
TRACE ELEMENTS IN COAL COMBUSTION[a]

Priority	Element or Ion	XRS	ES	SIE	AA/FP	FLUOR	COL
1	Fluoride			X[b]			
1	Lead	X	X		X		
1	Mercury	X	X		X		
1	Beryllium		X		X	X	
2	Cadmium	X	X		X		
2	Arsenic	X	X		X		
2	Nickel	X	X		X		
3	Copper	X	X		X		
3	Zinc	X	X		X		
3	Barium	X	X		X		
3	Tin	X	X		X		
3	Phosphorus						X
3	Lithium		X		X		
3	Vanadium	X	X		X		
3	Manganese	X	X		X		
3	Chromium	X	X		X		
3	Selenium	X	X		X		

[a]XRS - X-ray Spectrometry
ES - Emission Spectroscopy
SIE - Specific Ion Electrode Method
AA/FP - Atomic Absorption Spectrometry/Flame Photometry
FLUOR - Fluorimetry
COL - Colorimetric Method

[b]X - Probably applicable for analysis at trace concentration levels

Source: PB 237 366

number greater than about 15 (the atomic number of phosphorus), but it is
more sensitive for elements with higher atomic numbers. Generally, x-ray spectrometry is not sufficiently sensitive to be applicable at trace levels, particularly
if sample sizes are small. The x-ray spectrometer has the merit of internal stan-
dardization which neutralizes matrix effects and makes for relatively high accuracy. For a production system, the x-ray spectrometer should be coupled to
a computer.

An improved nondispersive x-ray spectrometer (NDXR) is particularly suited for
coupling to a computer because analysis of the fluorescent x-ray spectrum from
the sample is done by a solid-state detector/pulse-height analyzer that provides
digital data for computer processing.

Atomic absorption spectrometry is a method with high sensitivity, particularly
since new atomization procedures have been developed, for example, graphite
furnaces and plasma units. The method is suited for accurate analysis of specific
elements because (unlike emission and x-ray spectroscopy) it does not provide
information for more than two elements simultaneously. Atomic absorption is

particularly suited to the determination of mercury and would also be appropriate for lithium. It is most applicable to metals because the sensitivity of the method declines with increasing nonmetallic character of the element. The method is also less sensitive to elements that form refractory oxides (and hence atomize with greater difficulty).

The specific ion electrode method is preferred for measuring trace quantities of fluoride in aqueous solution. Generally, it is used with a pyrohydrolysis technique for liberating the fluoride as HF and collecting it in aqueous solution.

Fluorimetry is a very powerful method of analyzing very small quantities of certain elements. For example, it is the method of choice for uranium and beryllium. It is claimed that concentrations of beryllium as low as 10^{-5}% can be determined in air dust without chemical separations.

A variety of ultrasensitive colorimetric methods are available for many elements that are suited for trace element analysis of coal and coal combustion residues. Only phosphorus is listed here under the "Colorimetric" heading because other methods are more appropriate for the other elements. The colorimetric method is capable of measuring phosphorus in microgram quantities.

X-Ray Spectrometric Analysis of Coal, Limestone, and Coal Combustion Residues

X-ray spectrometry is a particularly attractive method for analyzing coal and coal combustion residues (CCCR) because [1] in principle, little sample preparation is required, [2] a number of elements are determined simultaneously, [3] application is to all elements of atomic number greater than 15, and [4] by the use of internal standards, quantitative information of relatively good accuracy can be acquired.

In X-ray spectrometry, as usually practiced, a sample of the material to be analyzed is irradiated with X-rays from a tungsten (or possibly molybdenum) target X-ray tube. Elements in the sample are excited and emit fluorescent X-rays whose energies are characteristic of the elements in the sample. The X-ray spectrum so produced is analyzed by a properly oriented crystal that disperses a beam of X-rays according to Bragg's law of diffraction. The intensities of the dispersed X-rays are measured by a radiation detector that physically traverses the area upon which they impinge. By suitable calibration, the intensities are converted to concentrations.

The X-ray spectrometer developed at Argonne National Laboratory (ANL) for fluidized bed combustion analyses differs from the conventional instrument in two respects: [1] the exciting radiation is itself a characteristic X-ray formed when a secondary target is irradiated with X-rays from a conventional X-ray tube, and [2] the detector is a solid-stage radiation detector/pulse-height analyzer that electronically sorts the fluorescent X-rays from the sample according to energy and measures the intensities.

The advantages of the new system are simultaneous measurement of all X-rays and markedly increased sensitivity. As little as tenths of a microgram of some elements can be detected. By adding known quantities of elements having about the same X-ray fluorescence energy as the element to be detected, the quantities of the elements present can be accurately estimated by direct comparison of the relative intensities.

Nondispersive X-Ray (NDXR) Spectrometry for Coal and Coal Combustion Residues

NDXR spectrometry can be used for quantitative simultaneous measurement of components at trace levels. When the concentrations of U, Au, Zn, and Fe in a sample were estimated to be 10, 0.1, 8-10, and 5 μg/ml of solution, respectively an analyzed sample consisted of less than 1 ml of solution with an internal standard. This instrument is useful to the analysis of CCCR.

If analysis for a particular element (or elements) were of interest, the sensitivity could be optimized by using a secondary target that would produce X-rays of an energy corresponding to the X-ray absorption edge of the element of interest. (This would enhance absorption and increase fluorescence X-ray intensity and, hence, sensitivity.)

FLUE-GAS SAMPLING

To determine the behavior of trace elements in coal during fluidized bed combustion, it is necessary to analyze the final flue gas for particulate and gaseous trace element compositions, as well as the various solids streams associated with the process. Flue-gas sampling procedures have been devised and the assembly of the necessary sampling equipment is described here.

Sampling Location

A very important consideration in trace element sampling is the location of the sampling point. Since the Brink Impactor used in the particulate sample handles particles in the range from 0.3 to 3 μm in diameter, the sampling is to be done after the flue gas has passed through gas-solid separators which remove particles larger than about 3 μm in diameter and through the pressure letdown valve. This has the advantage of allowing the sampling to be accomplished at near-atmospheric pressure, even though the combustor and the ancillary equipment are operating at pressures of up to 10 atm. The material removed in the separators can also be recovered and analyzed, if so desired. Modifications made in the flue-gas line to accommodate the sampling process are indicated in Figure 9.1.

During sampling periods, the sample-loop isolation valves are opened and the bypass valve is closed to direct the flue gas through the sampling zone. The sampling zone is a 5 ft length of 4 inch diameter Type L copper tube with a nominal inside diameter of 3.905 in. The increased cross section of the sampling zone is intended to reduce the average gas velocity past the sample probe to between 6 and 12 ft/sec, depending on the operating conditions in the combustor.

To allow for isokinetic sampling of the flue gas, the total gas flow through the sample loop will be determined using the orifice meter. Based on the relative conditions at the meter and the sampling location, the linear gas-flow rate past the sample probe can be determined because there is fully developed flow at the sampling point, since the sample probe is located approximately 12 pipe-diameters downstream and 3 pipe-diameters upstream from any flow disturbances.

FIGURE 9.1: EXPERIMENTAL FLOW SYSTEM FOR FLUE-GAS SAMPLING

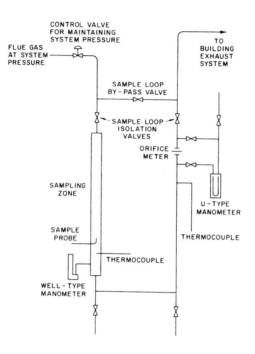

Source: PB 237 366

Particulate Sampling

The apparatus for the sampling of flue-gas particles is illustrated in Figure 9.2. The sample probe is constructed from stainless steel tubing and a tapered nozzle is affixed to the end of the probe to provide for streamlined sampling. The probe can be inserted to any depth along the diameter of the flue to allow for sampling at several traverse points. The probe, which normally faces upstream, can also be pointed downstream when samples are not being taken to prevent possible accidental plugging of the nozzle opening.

The particles in the gas sample are collected in the Brink Cascade Impactor and the follow-up, glass-fiber filter (Gelman Type A). The impactor has five in-line stages, each of which has a jet that utilizes a collection cup as an impaction plate. A spring holds each collection cup in place. A single stage of the impactor is illustrated in Figure 9.3.

The particles suspended in the gases pass through a jet. Particles with sufficient inertia impact against a cup, and the remainder pass through annular slots located around the cup.

Each collection cup has annular slots with a total cross-sectional area 30 times the area of the largest jet, thereby reducing turbulent effects to a negligible level.

FIGURE 9.2: APPARATUS FOR PARTICULATE SAMPLING OF FLUE GAS

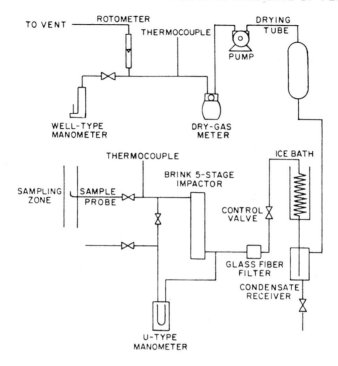

FIGURE 9.3: SINGLE IMPACTOR STAGE

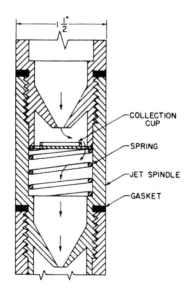

Source: PB 237 366

The impactor collects particles in the range from 0.3 to 3.0 μm. The glass-fiber filter at the outlet of the impactor collects the particles that pass the final stage; in general, these are particles smaller than 3 μm.

The control valve is used to regulate the volumetric flow through the impactor from 0.9 to 1.2 cfm, so that the velocity at the nozzle inlet of the sample probe equals the linear velocity of the flue gas flowing past the probe. This is done to promote representative sampling of the particles in the flue gas by isokinetic sampling. Heating the probe, impactor, and filter may be necessary to prevent condensation of flue-gas water.

The remainder of the sample train is designed to accurately determine the sampling rate and the total volume of the gas sample taken, so that the particulate loading of the main flue-gas stream can be calculated.

Treatment of Particulate Samples

In addition to obtaining total particulate and trace element concentrations in the flue gas, it should also be possible to obtain trace element data as a function of particle size. It is necessary, therefore, to obtain complete analytical and particle-size-distribution data for the material impacted on each cup or stage of the impactor. Since it is not uncommon to collect from 40 to 80% of the particulate matter on one stage of the impactor, the sample time will be limited by reentrainment from this cup. The maximum sample time must be used, therefore, to insure the collection of reasonable quantities of material on the other cups.

To obtain particle-size-distribution data, samples of the material from each stage and the final filter are to be examined using a scanning electron microscope (SEM). Reasonably good photomicrographs can be obtained at magnifications as high as 20,000X; it should be possible, therefore, to obtain from such photographs reasonably accurate particle-size information. To obtain the necessary elemental information, the material from each stage of the impactor will also be subjected to chemical analysis.

Mercury Determination in Particulate and Gaseous Emissions

Supplemental to the primary determination of particulate emissions in the flue gas is the determination of possible trace element gaseous emissions. Of particular concern in this area is mercury since up to 90% of the mercury in coal is released during combustion and appears as vapor discharged in the stack gas, while only 10% remains in the residual ash. It is essential, therefore, that for mercury, a method for sampling both particulate and gaseous emissions needs to be used.

The Environmental Protection Agency has proposed an apparatus to be used in the sampling train as illustrated in Figure 9.4. In principle, the particulate and gaseous emissions are isokinetically sampled from the flue gas and collected in the acidic iodine monochloride solution. The mercury collected (in the mercuric state) is reduced to elemental mercury in basic solution by hydroxylamine sulfate. Mercury is subsequently vaporized from the solution by a mercury-free air stream and analyzed using an atomic absorption spectrophotometer in the flameless mode.

FIGURE 9.4: APPARATUS FOR THE DETERMINATION OF PARTICULATE
AND GASEOUS EMISSIONS OF MERCURY IN FLUE GAS

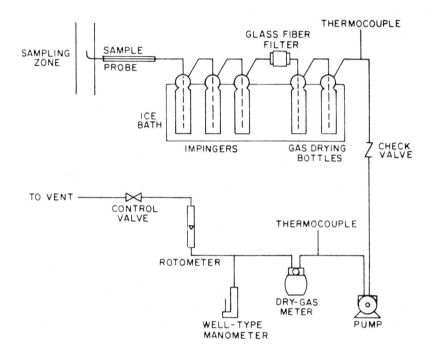

Source: PB 237 366

The entire sampling train up to the check valve is Pyrex glass. The probe dimensions are to be somewhat larger than the stainless steel probe to accommodate the larger volumetric sampling flow rates of 0.5 to 1.0 cfm.

The second impinger in the sampling train is of the Greenburg-Smith design with the standard tip. The remainder of the impingers are gas-washing bottles modified by replacing the standard tip with a ½ inch ID glass tube, which extends to ½ inch from the bottom of the flask. Each flask has a capacity of 500 ml. The first, second, and third impingers contain the mercury scrubbing solution. The fourth impinger contains silica gel to remove water vapor.

The filter holder is also Pyrex and designed to accept 50 mm glass-fiber filter paper. The filter provides for the removal of particulates and liquid mist entrained in the flowing gas from the first three impingers.

Determination of Inorganic Fluorides in Gaseous and Particulate Emissions

It is also possible that some fluorine may be present in the gaseous phase of the flue gas from the combustor. Gaseous and particulate inorganic fluorides can be sampled from the atmosphere using a water-filled impinger. It is possible, therefore, to use the same sampling train for both the mercury determination and the fluoride determination.

TRACE ELEMENT DISTRIBUTION STUDIES

Mercury Mass Balances

Mercury balances for four experiments are presented in Table 9.3. Experiments 1 and 2, which were completed first (with two ICl scrubbers in the gas-sampling apparatus), exhibited Hg recoveries, expressed as the percentage of Hg entering the combustor which can be accounted for in the combustion products, of 56 and 43%, respectively.

For Experiments 3 and 4 (with two gold frits and one ICl scrubber in the gas sampling apparatus) the respective recoveries were only 29 and 25%. The lower total Hg recovery values for the latter experiments apparently resulted from the decreased recovery of Hg in the volatile form, which was lower by a factor of 6 as compared with the earlier experiments (3 and 5% vs 19 and 34%). The amount of Hg recovered in the solid-phase combustion products ranged from 9 to 37% with an average recovery of 23% for the four experiments.

Measured concentrations of Hg in the flue gas for Experiments 3 and 4 (made at 1650°F) were only 0.03 and 0.05 ppb as compared with 0.32 and 0.66 ppb for Experiments 1 and 2 (made at 1550°F). It would seem more logical to expect an increase in the level of Hg in the flue gas at the higher combustion temperatures. The lower measured concentrations do not, however, appear to result from the modifications in the flue gas sampling equipment as discussed.

In experiments with the modified apparatus, traces of Hg were found in the carbonate solution (considerably less than in the first two experiments, however), on the gold frits, and on the filter following the ICl scrubbing solution. As with the first two experiments, no Hg was detected in the ICl solution.

Retention of Mercury in Solid Products of Combustion

It has been found that mercury is recovered in solid combustion products. A mercury balance has been made around a 660 MWe coal-fired power station using a coal containing 21% ash (<1% S) and having a heating value of approximately 9000 Btu/lb. Approximately 10% of the mercury remained with the furnace residual ash, and 90% was emitted in the vapor phase. The average concentrations of Hg in the coal and ash products were 0.3 ppm and 0.2 ppm, respectively.

The fly-ash samples in the above study were less concentrated in Hg than the coal being combusted, whereas in the experiments just discussed the elutriated solids were consistently higher in Hg concentration than the coal (except for primary cyclone samples highly diluted with sulfated dolomite). This is reflected in the higher retention of Hg by the ash products (37 and 26%) reported in Table 9.3. This is an indication, at least, that fluidized bed combustion may offer the potential for increased retention of mercury by the solid products of combustion.

TABLE 9.3: MERCURY MATERIAL BALANCES FOR TRACE ELEMENT COMBUSTION EXPERIMENTS

Component	Combustion in Alundum Bed				Combustion in Dolomite Bed			
	Experiment 1 (1550°F)[a]		Experiment 3 (1650°F)		Experiment 2 (1550°F)		Experiment 4 (1650°F)	
	Conc., ppm	Wt, µg	Conc., ppm	Wt, µg	Conc., ppm	Wt, µg	Conc., ppm	Wt, µg
IN: Coal	0.15	5,900	0.07	5,680	0.15	9,100	0.07	7,050
Additive	—— No Additive Used ——				0.017	510	0.025	530
Initial Bed	<0.005[b]	Neg.[c]	0.028	640	<0.01	Neg.	0.005	70
TOTAL Hg In		5,900		6,300		9,600		7,600
OUT: Final Bed	<0.005	Neg.	<0.005	Neg.	<0.01	Neg.	<0.005	Neg.
Overflow	—— No Overflow ——				<0.01	Neg.	<0.005	Neg.
Prim. Cyclone	0.46	2,100	0.15	1,280	0.035	520	0.06	990
Sec. Cyclone	0.46	130	0.39	130	0.28	170	0.35	420
Prim. Filter	—— No Sample ——		0.45	150	0.15	90	0.28	110
Sec. Filter	—— No Sample ——		0.52	10	—— No Sample ——		—— No Sample ——	
Flue Gas	0.32 ppb	1,100	0.03 ppb	180	0.66 ppb	3,300	0.05 ppb	390
TOTAL Hg Out		3,300		1,800		4,100		1,900
Recovery,[d] %:								
TOTAL		56		29		43		25
In Flue Gas		19		3		34		5
In Solid Form		37		26		9		20

[a] Bed temperature
[b] Estimated
[c] Considered negligible
[d] Recovery = (Hg out/Hg in) × 100%

Source: PB 237 754

Studies have also been made of Hg during the combustion of coal in combustion units of various sizes. In a 100 g/hr test unit, approximately 60% of the mercury was retained in the fly ash (combustion at 2100°F) when burning a coal containing 0.15 ppm Hg (approximately 0.9 ppm in the ash). Approximately, 35% of the mercury was retained in the fly ash when burning a coal containing 0.24 ppm Hg (approximately 0.35 ppm in the ash).

In a 500 lb/hr combustor burning pulverized coal containing 0.18 ppm Hg, ash removed in cyclones at approximately 360°F contained 0.22 ppm Hg, which accounted for approximately 12% of the Hg in the coal. Samples of fly ash from 250 MWe and 175 MWe power plants indicated mercury contents in the ash of 0.10 and 0.26 ppm, respectively, for estimated Hg retentions of 7 and 19% in the fly ash.

Lead and Beryllium Mass Balances

Lead balances for four trace element experiments are presented in Table 9.4. The material balances range from 78 to 125% recovery of the lead entering the combustor. Beryllium balances for the trace element experiments are presented in Table 9.5. Recoveries of Be were somewhat lower than for the Pb, ranging from 56 to 87%.

The results strongly indicate that the lead, with the observed high recoveries, is essentially retained by the particulate matter in the combustion products. Emissions of Pb, therefore, will be controlled by the efficiency of particulate removal from the flue gas. However, because the concentration of Pb in particulate matter increases with decreasing particle size, the efficiency of lead removal will be somewhat lower than the overall efficiency of particulate removal.

The beryllium mass balances, with the lower recoveries, suggest the possible volatilization of beryllium (or compounds of beryllium) during combustion although this seems an unlikely possibility for the bed temperatures used in these combustion experiments.

Preferential Concentration of Lead and Beryllium in Finer Particulate Matter

The results indicate that both lead and beryllium concentrate preferentially in the finer particles leaving the combustor. The combustion gases leaving the combustor pass through a primary cyclone, a secondary cyclone, and a porous filter, in that order. Particle-size analysis curves for the primary and secondary cyclone materials as recovered from Experiment 1 indicate a 50% cutoff value of approximately 10 μm for the "mean" diameter of the primary cyclone material.

A mean value of 1.5 μm was found for the diameter of the secondary cyclone material. For the coarser primary-cyclone ash product, the lead and beryllium concentrations were 95 and 2.65 ppm, respectively. In the finer, secondary-cyclone ash, the lead and beryllium concentrations increased to 255 and 5.94 ppm, respectively.

The preferential concentration of relatively nonvolatile trace elements in particles of smaller diameter has been demonstrated in samples taken from ambient air (1)(2) and in fly-ash particles retained in the precipitation systems and in the airborne fly ash leaving eight different coal-fired power plants in the United States (3).

TABLE 9.4: LEAD MATERIAL BALANCES FOR TRACE ELEMENT COMBUSTION EXPERIMENTS

| | Combustion in Alundum Bed | | | | Combustion in Dolomite Bed | | | |
| | Experiment 1 (1550°F)[a] | | Experiment 3 (1650°F) | | Experiment 2 (1550°F) | | Experiment 4 (1650°F) | |
Component	Conc., ppm	Wt, mg	Conc., ppm	Wt, mg	Conc., ppm	Wt, mg	Conc., ppm	Wt, mg
IN: Coal	29	1,100	1.6	130	29	1,800	1.6	160
Additive	---- No Additive Used ----				6	180	12	310
Initial Bed	1.1	14	1.1	25	21	290	11	150
TOTAL Pb In		1,100		160		2,300		620
OUT: Final Bed	51	670	2.8	60	16	260	11	150
Overflow	----- No Overflow -----				15	220	12	100
Prim. Cyclone	95	440	15	130	70	1,000	16	270
Sec. Cyclone	260	70	13	4.3	180	110	27	33
Prim. Filter	- No Sample -		22	7.4	300	170	96	38
Sec. Filter	- No Sample -		46	0.6	-------- No Sample --------			
TOTAL Pb Out		1,200		200		1,800		590
Recovery, %: TOTAL		109		125		78		95

[a]Bed temperature.

Source: PB 237 754

TABLE 9.5: BERYLLIUM MATERIAL BALANCES FOR TRACE ELEMENT COMBUSTION EXPERIMENTS

| | Combustion in Alundum Bed | | | | Combustion in Dolomite Bed | | | |
| | Experiment 1 (1550°F)[a] | | Experiment 3 (1650°F) | | Experiment 2 (1550°F) | | Experiment 4 (1650°F) | |
Component	Conc., ppm	Wt, mg	Conc., ppm	Wt, mg	Conc., ppm	Wt, mg	Conc., ppm	Wt, mg
IN: Coal	0.7	28	0.66	54	0.7	42	0.66	66
Additive	----- No Additive Used -----				0.67	20	0.75	16
Initial Bed	0.83	10	0.83	19	0.78	11	0.66	8.9
TOTAL Be In		38		73		73		91
OUT: Final Bed	0.76	10	0.79	17	0.75	12	0.80	11
Overflow	----- No Overflow -----				0.73	10	2.44	21
Prim. Cyclone	2.65	12	2.29	19	1.55	23	2.24	37
Sec. Cyclone	5.95	2	6.62	2.2	5.20	3	5.63	6.8
Prim. Filter	– No Sample –		6.75	2.3	6.77	4	7.70	3.0
Sec. Filter	– No Sample –		8.05	0.1	----- No Sample -----		-----	
TOTAL Be Out		24		41		52		79
Recovery, %: TOTAL		63		56		71		87

[a]Bed temperature.

Source: PB 237 754

It has been proposed that preferential concentration occurs by volatilization of the elements (or one of their compounds) during combustion, followed by condensation or adsorption onto the larger surface area, per unit mass, of the smaller particles (3). It was noted that the normal temperature in the combustion zones of the conventional plants tested was between 1300° and 1600°C, a temperature similar to or above the boiling points of the elements investigated (Cd, As, Ni, Pb, Cr, and Zn).

Although the test results described here appear to confirm the preferential concentration of Pb and Be in smaller particles, there is little evidence to support the idea of volatilization and subsequent condensation or absorption as the mechanism by which this concentration occurs.

In a report by the Illinois Geological Survey, it was shown that there was no significant loss of beryllium in coal ash prepared at 700°C (4). As Experiments 1 and 2 were carried out at 10 atm pressure and a bed temperature of only 850°C, it does not appear likely that volatilization would be a significant factor in the apparent concentration of the beryllium in the finer particles. A somewhat simpler explanation may be that the trace elements in the larger particles are simply diluted by unburned carbonaceous materials. This is evidenced in Table 9.6, which tabulates the concentrations of trace, minor, and major elements in particulate matter recovered during the experiments at various stages of removal from the flue gas.

As Experiments 1 and 3 were made using an Alundum bed, the changes in concentration with subsequent stages of solids recovery from the flue gas are not biased by the presence of large amounts of additive. In Experiment 2, the level of carbon decreases from 51.3 to 29.2% as the concentrations of the trace and minor elements in the coal increase in the ash. In Experiment 3, the concentration of carbon decreases from 34.1 to 6.6% as the levels of the other elements show concentration increases.

An interesting observation can also be made by comparing the ratios of lead to beryllium in the coal, additive, and in each of the materials removed at various stages of the gas cleanup system. These ratios are presented in Table 9.7.

The columns of particular interest are those under Experiments 1 and 3, in which additive does not appear as a factor in the calculated ratios. While there does appear to be some variation in the ratio of Pb to Be, these variations could well fall within the accuracies of the analytical methods. If volatilization and condensation are the mechanisms by which preferential concentration occurs, it seems likely that the Pb-to-Be ratio would change significantly over the range of samples analyzed.

Fluoride Mass Balances

Fluoride material balances for the trace element experiments are given in Table 9.8. The indicated recovery of fluoride for the experiments at 1550°F are 123 and 110%, which are reasonably acceptable values. The recoveries of 180 and 240 reported for the experiments at 1650°F, however, are unaccountably high. The only differences in sampling between the two sets of experiments were the use of considerably larger flue-gas samples and the use of Na_2CO_3 scrubbers instead of $NaHCO_3$ scrubbers for the experiments at 1650°F.

TABLE 9.6: CONCENTRATIONS OF TRACE, MINOR, AND MAJOR
ELEMENTS IN RECOVERED PARTICULATE MATTER

Source of Solids	Concentration				
	Pb, ppm	Be, ppm	C, wt %	S, wt %	Ca, wt %
Experiment 1					
Primary Cyclone	95	2.65	51.3	0.86	1.29
Secondary Cyclone	260	5.95	29.2	1.80	1.80
Experiment 2					
Primary Cyclone	70	1.55	2.3	4.5	15.4
Secondary Cyclone	180	5.20	13.1	2.6	5.58
Filter	300	6.77	8.4	3.1	6.65
Experiment 3					
Primary Cyclone	15	2.29	34.1	1.8	2.11
Secondary Cyclone	13	6.62	16.8	2.5	2.02
Primary Filter	22	6.75	6.6	4.9	2.89
Secondary Filter	46	8.05	----	----	----
Experiment 4					
Primary Cyclone	16	2.24	19.6	4.8	8.11
Secondary Cyclone	27	5.63	11.3	3.1	4.89
Primary Filter	96	7.70	5.2	4.3	5.44

TABLE 9.7: RATIO OF LEAD TO BERYLLIUM IN RAW MATERIALS
AND PARTICULATE MATTER

Material	Experiment 1	Experiment 2	Experiment 3	Experiment 4
Coal	41	41	2.4	2.4
Additive	not used	9	not used	16
Primary Cyclone	36	45	6.5	7.1
Secondary Cyclone	43	34	2.0	4.8
Primary Filter	no sample	44	3.3	12
Secondary Filter	no sample	no sample	5.7	no sample

Source: PB 237 754

TABLE 9.8: FLUORIDE MATERIAL BALANCES FOR TRACE ELEMENT COMBUSTION EXPERIMENTS

| | Combustion in Alundum Bed | | | | Combustion in Dolomite Bed | | | |
| | Experiment 1 (1550°F)[a] | | Experiment 3 (1650°F) | | Experiment 2 (1550°F) | | Experiment 4 (1650°F) | |
Component	Conc., ppm	Wt, g	Conc., ppm	Wt, g	Conc., ppm	Wt, g	Conc., ppm	Wt, g
IN: Coal	25	1.0	29	2.4	25	1.5	29	2.9
Additive	No Additive Used				14	0.4	14	0.3
Initial Bed	100	1.2	100	2.3	350	4.9	130	1.8
Total F⁻ In		2.2		4.7		6.8		5.0
OUT: Final Bed	Not Detected	-	36	0.8	38	0.6	86	1.2
Overflow	No Overflow				86	1.2	41	0.4
Prim. Cyclone	20	0.1	36	0.3	150	2.3	71	1.2
Sec. Cyclone	10	Neg.	12	Neg.	47	Neg.	37	Neg.
Prim. Filter	No Sample	-	5	Neg.	115	Neg.	29	Neg.
Sec. Filter	No Sample	-	Insufficient Sample		No Sample	-	No Sample	-
Flue Gas		2.6		7.4		3.3		9.2
Total F⁻ Out		2.7		8.5		7.4		12.0
Recovery, %: TOTAL	123		180		110		240	
In Flue Gas	118		157		48		184	
In Solids	5		23		62		56	

[a]Bed temperature.

Source: PB 237 754

Perhaps the most significant observation that can be made from the F⁻ balances is that the retention of the F⁻ in the solid phases leaving the combustor appeared to be significantly higher when additive was used in the experiment. The reported recoveries of F⁻ in the solid samples were 56 and 62% for combustion with additive present and only 23 and 5% for combustion in an Alundum bed. A possible explanation for such a phenomenon could be the formation of CaF, which is a relatively stable compound.

Neutron Activation Analysis

Neutron activation analysis is an instrumental method for expanding the trace element study to include elements of second- and third-priority interest (Cd, As, Ni, Zn, Cu, Ba, Sn, P, Li, Mn, Cr, Se, and V). A preliminary testing of the method was made by taking 100 mg samples of coal, primary- and secondary-cyclone ash, and irradiating them in a test reactor for 2.5 hours.

After the irradiation, periodic γ-ray counts were taken using a Ge(Li) detector on all three samples for the purpose of identification of activation products and relative activity levels. Results for seven of the elements detected are presented in Table 9.9; additional elements were also detected, but are not listed. With the exception of the material balance for iron, the remaining four elements balance as well as, or better than, the values obtained for mercury, lead, beryllium and fluoride by direct chemical analysis.

TABLE 9.9: NEUTRON ACTIVATION RESULTS ON SAMPLES FROM TRACE ELEMENT EXPERIMENT

| Element | Concentration, ppm | | | Material Balance, % Recovery |
	Coal	Primary Cyclone	Secondary Cyclone	
Fe	11,300	33,000	39,000	37
Cr	--	250	330	--
Sc	2.4	18	23	96
Na	960	11,000	10,000	143
K	460	2,800	2,800	76
Zn	710	1,600	--	--
Cu	33	330	410	127

Source: PB 237 754

Sodium Concentrations in Flue Gas Particulates

Samples from several experiments have been analyzed for Na because of the considerable interest that has been expressed in Na as a source of corrosion in a combined-cycle power system, particularly corrosion of the blades of a high-temperature, gas turbine. The results of several analyses are presented in Table 9.10.

FIGURE 9.10: SODIUM MATERIAL BALANCES FOR TRACE ELEMENT COMBUSTION EXPERIMENTS

Component	Combustion in Alundum Bed				Combustion in Dolomite	
	Experiment 1		Experiment 3		Experiment 4	
	Conc.,[a] wt %	Wt, g	Conc., wt %	Wt, g	Conc., wt %	Wt, g
IN:						
Coal	0.10[c]	40	0.05[b]	41	0.05	50
Additive	--- No Additive Used ---				0.04	8.4
Initial Bed	0.11	14	0.11	25	0.06	8.1
Total Na In		54		66		66
OUT: Final Bed	0.12	16	0.06	13	0.07	10
Overflow	--- No Overflow ---				0.06	5.2
Prim. Cyclone	1.1[c]	51	0.47	40	0.35	57
Sec. Cyclone	1.0[c]	2.9	0.88	2.7	0.95	11
Prim. Filter	--- No Sample ---		1.97	6.6	1.51	6
Sec. Filter	--- No Sample ---		1.65	0.2	--- No Sample ---	
Total Na Out		70		62		89
Recovery, % TOTAL		130		94		135

[a]Determined by atomic absorption spectroscopy.

[b]Commercial Testing Company, Chicago, Illinois reported a value of 0.07 wt % Na in Arkwright coal.

[c]Determined by neutron activation analysis.

Source: PB 237 754

It would appear from the material balances that the Na is retained by the particulate matter during combustion. As with Pb and Be, the concentration of Na in the particulate matter increases with decreasing particle size. Concentrations generally vary from 0.5 weight percent for material removed in the primary cyclone to 1.5-2.0 weight percent for material removed in the primary and secondary filters.

It should be emphasized however, that the flue gas is cooled to approximately 600° to 800°F during the solids removal process. Because the flue gas entering the turbine is at considerably higher temperatures (approximately 1600°F), considerably less Na may be retained by the particulates during their removal at these higher temperatures.

The samples which had been analyzed for sodium were also analyzed for chloride and sulfate. It was concluded that only a small amount of sodium was present as NaCl, whereas a large quantity was present as Na_2SO_4.

EFFECT OF TRACE EMISSIONS ON GAS-TURBINE PERFORMANCE

Alkali-Metal Compound Effects

Sodium and potassium compounds are potentially hazardous to the operation of the gas turbine. Chlorides and hydroxides are volatile species and can transport sodium and potassium from the combustor to the turbine. At hydrogen chloride (HCl) levels exceeding 0.4 ppm by volume in the combustor gas, solid or liquid sodium sulfate (Na_2SO_4) will convert to gaseous sodium chloride (NaCl).

The hydrogen chloride level in the combustion gas resulting from the complete release of chlorine from a low-chlorine coal (100 ppm Cl) exceeds this level by over a factor of 10 and is 5 ppm. In a fluid bed combustion process the predominant transport should be by the chlorides.

In the gas turbine, reactions between the chloride and the sulfur oxides in the combustion gas will form liquid sulfate-chloride melts on the turbine hardware if the sodium and potassium levels are sufficiently high. These melts must be prevented because they initiate hot-corrosion and deposit formation. Hydrogen chloride in the turbine acts to prevent sulfate deposits from forming or, once formed, acts to remove them.

A hydrogen chloride level of 40 ppm by volume in the combustion gas is sufficient to prevent a liquid sodium sulfate melt from being stable at sodium concentrations in the gas up to 0.2 ppm by volume. (Forty ppm hydrogen chloride corresponds to complete release of chlorine from coal containing 800 ppm by weight of chlorine; 0.2 ppm of sodium corresponds to a 1% release of sodium from a coal containing 130 ppm by weight of sodium.)

The concentration both of hydrogen chloride and of sulfur oxides (SO_2 and SO_3) has a strong influence on the stability of the melt. If the sulfur dioxide level in the gas is dropped from 200 to 100 ppm by volume, the concentration of hydrogen chloride required to prevent deposition of a liquid sodium sulfate film would drop to about 25 ppm by volume.

Sodium and Potassium Tolerances

Three factors must be considered for defining the sodium and potassium tolerances which will prevent hot corrosion attack in the turbine. These are:

[1]   The influence of the interaction between sodium and potassium to form complex melts on the hardware. In such melts the activities of the sodium and potassium are reduced, and the equilibrium concentrations of sodium and potassium species that can exist in gas above the melt are also lowered. Interaction tends to reduce the tolerable concentrations of sodium and potassium in the turbine expansion gas. Studies to establish the magnitude of this effect and also to establish the influence of the relative sodium and potassium levels on the composition and melting point of stable deposits are being carried out.

[2]   The ability of turbine stator vane and rotor blade alloys to withstand a combustion gas containing up to 200 ppm sulfur dioxide and up to 40 ppm hydrogen chloride.

[3]   The shifts in the turbine tolerance which will occur if equilibrium levels of sulfur trioxide are not achieved and the degree to which kinetic factors such as these influence the tolerable concentration of sodium and potassium chlorides.

In the combustor itself a competition takes place for the sodium and the potassium. Hydrogen chloride attacks the sodium and potassium compounds attempting to form the volatile chlorides. On the other hand, in the burning char particle, silica reacts with sodium to form stable silicates. Unfortunately, potassium silicates are not stable, and potassium clay mineral compounds are breaking down and converting to either polysulfides or chlorides, depending on the composition of the local atmosphere.

In the oxidizing atmospheres outside of combusting particles, the sulfates are stable at low levels of hydrogen chloride; but at levels above 0.4 ppm, reconversion of condensed sulfates to chlorides will occur. Attempts are being made to establish the feasibility of controlling the alkali metal content of the gas through:

●   Control of the level of hydrogen chloride in the combustor. Reducing hydrogen chloride levels in the combustor below 0.4 ppm by volume would lower the sodium chloride content of the transport gas by slowing or preventing attack of ash minerals and by promoting conversion to low-volatility condensed sulfate.

●   Conversion of high-volatility sodium and potassium chlorides in coal and dolomites to sulfates or silicates, with simultaneous removal of hydrogen chloride by low-temperature pretreating in the coal and stone dryers and preheaters.

REFERENCES

(1)   R.E. Lee, Jr., S.S. Goranson, R.E. Enrione, and G.B. Morgan, "National Air Surveillance Cascade Impactor Network II. Size Distribution Measurements of Trace Metal Components," *Environ Sci and Technol 6,* 1025 (1972).

(2)   M. Kertesz-Saringer, E. Meszaros, and T. Varkoni, *Atmospheric Environment 5,* 429 (1971).

(3)   D.F.S. Natusch, J.R. Wallace, and C.A. Evans, Jr., "Toxic Trace Elements: Preferential Concentration in Respirable Particles," *Science 183,* 202 (Jan. 18, 1974).

(4)   R.R. Ruch, H.J. Gluskoter, and N.F. Shimp, *Occurrence and Distribution of Potentially Volatile Trace Elements in Coal,* Illinois State Geological Survey, Environmental Geology Notes No. 61 (1973).

Part II.
Waste Handling in Fluidized Beds

The inherent characteristics of fluidized combustion make it ideally suited to the burning of sewage sludge, oily wastes, refinery sludges and municipal waste. In fact, any fuel or mixture of fuels can be burned in a fluidized bed, be it gas, oil or solids.

RESIDUAL OIL
GASIFICATION/DESULFURIZATION

The reports used in this chapter were excerpted from: PB 212 960; PB 233 101; PB 241 834; PB 241 835; and *Technology Ireland*. For a complete bibliography, see p 263.

OBJECTIVES OF USING RESIDUAL OIL

Achievement of the national goal of fuel resource independence rests largely on the success of massive development programs in coal and nuclear fuel utilization. It is clear that petroleum derivatives will continue to supply a large portion of the United States' electric utility fuel demand during the years before national energy independence is realized. Two options presently exist for the environmentally sound utilization of atmospheric residual fuel oils for electric power generation: conversion to low-sulfur fuel oils (hydrodesulfurization – HDS) at the refinery or stack-gas cleaning (limestone slurry scrubbing, lime slurry scrubbing, and others) as a utility boiler retrofit or new boiler feature.

In most cases the utilization of refinery vacuum bottoms as a fuel oil is not economically feasible with these two options because of their high metals content. With the FB residual oil gasification process the goal is to provide a third alternative which utilizes heavy fuel oils (atmospheric residual or vacuum bottoms) to generate clean power in terms of gaseous emissions and liquid and solid wastes.

The EPA through ORD is funding the development of an atmospheric-pressure fluidized bed oil gasification process for firing in a conventional power plant. A simplified illustration of the process concept is shown in Figure 10.1.

The Esso Petroleum Company, Ltd., Abingdon, England (Esso) invented the chemically active fluidized bed (CAFB) process and carried out pilot-plant studies. Residual fuel oil is added to a simple, open, fluidized bed of limestone with sufficient air, to 25% stoichiometric, to maintain the bed at about 870° to 970°C (1600° to 1778°F) and to react the oil to produce a hot fuel gas. The fuel oil is partially burned in the refractory-lined gasifier vessel by a mixture of air and recirculated steam generator flue gas, to maintain temperature of the reaction.

FIGURE 10.1: ATMOSPHERIC FLUIDIZED BED OIL GASIFICATION POWER PLANT

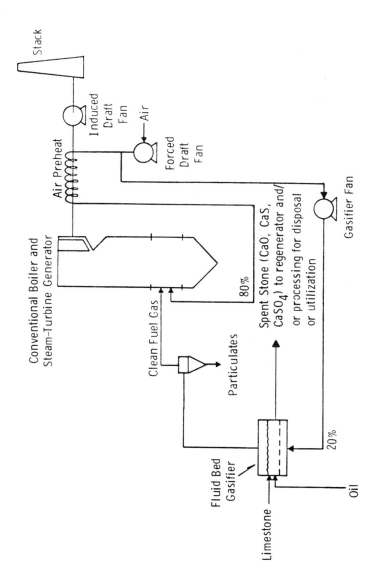

Source: PB 241 834

By maintaining the proper Ca/S ratio again within the gasifier, greater than 90% of fuel sulfur is captured. The product gas from the gasifier vessel has a heating value of 200 Btu/cu ft and is transported through a dust collector to conventional low Btu gas burners.

The calcium sulfite formed in the gasifier is injected into an adjacent regenerator zone where it is fluidized with air. The heat of reaction of the calcium sulfide-to-calcium oxide and SO_2 plus the carbon deposits on the particles, provide enough energy to maintain the regenerator temperature. The regenerated calcium oxide is then returned to the gasifier.

The off-gas from the regeneration process contains 8 to 10% SO_2 by volume but the flow of this SO_2-rich off-gas is about 3% of the flow rate at the steam generator stack. A significantly smaller sulfur recovery system is required to treat the total steam generator flue gas stream. It may also be possible to burn noncaking coals in this way. CAFB gasifiers are expected to become commercially available by early 1980s.

A summary of the Esso development work up until the beginning of 1975 is shown in Table 10.1. With EPA funding Esso operated a 750 kW CAFB pilot plant for more than 2,600 hours and provided design and operating information required for the development program.

TABLE 10.1: PHASES OF CAFB EXPERIMENTAL DEVELOPMENT

Phase	Accomplishments	Time period
Pre-EPA	Feasibility in batch units	1966-1970
EPA Phase I	• Comparison of fuels and limestones in batch units • Intensive process variables study in batch units • Design, construction, and commissioning of continuous pilot plant	June 1970-March 1972
EPA Phase II	• Comparison of fuels and limestones in batch units • Study of process variables, design alternatives, and process problems in continuous pilot plant	July 1972-Dec. 1973
EPA Phase III	• Support for demonstration plant design • Investigation into trace element retention	July 1973-Jan. 1975

Source: PB 241 834

In November 1972, New England Power Service Company of New England Electric System (NEES) agreed to participate in the program, and boiler No. 12

at the Manchester Street Station, Narragansett Electric Company, Providence, R.I. was selected as the demonstration plant site. A three-phase demonstration plant program was conceived:

(1) Phase 1 — Preliminary design and cost estimate of demonstration plant installed on an existing boiler, followed by an assessment by all parties and a decision to stop or proceed;

(2) Phase 2 — Detailed design and construction of demonstration residual oil gasification process; and

(3) Phase 3 — Developmental operation of the gasification process and integrated power plant.

Westinghouse, as prime contractor to EPA, subcontracted with Stone & Webster Engineering Corporation (SWEC) for the engineering services for Phase 1. Combustion Engineering, Inc. (CE) was engaged in a boiler retrofit and burner evaluation study. Objectives for the work performed under this contract included:

(1) Assessment of the market for the process;

(2) Selection of complete, integrated fluidized bed oil gasification/desulfurization/power plant concept;

(3) Preparation of a preliminary design and cost estimate for the 50 MW demonstration plant;

(4) Preparation of a conceptual design and cost estimate for a 200 MW plant;

(5) Specification of a work scope required to carry out the demonstration plant program; and

(6) Assessment of alternative technology in light of the prior tasks.

GASIFICATION/DESULFURIZATION CONCEPTS

Two possible modes for gasification/desulfurization operation are the regenerative mode and the once-through mode. Figure 10.2 illustrates the major process streams and identifies the basic elements of the two operational modes, without reference to the specific system configuration or specific retrofit concepts. The operating variables and temperature control schemes differ slightly for the two modes.

The basic elements of the regenerative operation are the gasifier vessel and the regenerator vessel, as shown in Figure 10.2. In the regenerative operation, residual oil is injected into the gasifier vessel, an air-fluidized bed of lime at 1600°F operated with substoichiometric air ($\sim$ 20% of stoichiometric), to yield by cracking, partial combustion, and H_2S absorption by the lime, a hot, low-sulfur, fuel gas and sulfided lime.

The fuel gas is transported to the boiler burners, where combustion is completed, and the sulfided lime is sent to the regenerator. The regenerator is an air-fluidized vessel operated with a slight excess of air at 1900°F. Regeneration takes place by reaction of oxygen with the utilized lime to give an SO_2-rich stream (of about 10 mol % SO_2) and a regenerated lime with a decreased activity compared to that of fresh lime.

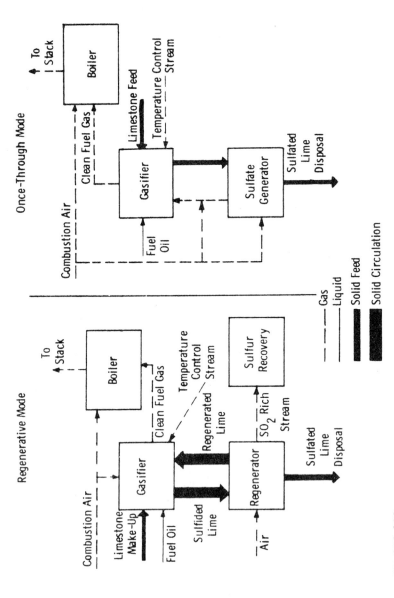

FIGURE 10.2: MODES OF OPERATION

Source: PB 212 960

The SO_2 stream is transported to a sulfur recovery system, and the regenerated lime is returned to the gasifier along with a nearly stoichiometric amount of fresh make-up limestone.

A variety of temperature control schemes is useful in controlling the temperatures of the gasifier and the regenerator. The temperature control scheme used in the gasifier must conserve the energy of the fuel in a form usable in the boiler. For this reason water injection, steam injection, and stack gas recycle are considered for temperature control of the gasifier by the storage of sensible heat; while the use of heat transfer surface in the gasifier to preheat boiler water is also an attractive possibility. A fifth possibility is to reduce the air/fuel ratio to a value low enough ($\sim$ 14% of stoichiometric) that the gasifier is thermally balanced and no further temperature control need be considered.

Because of the high cost of sulfur recovery, temperature control of the regenerator is best accomplished in such a way as to avoid diluting the SO_2 stream produced by regeneration. For this reason temperature control by heat transfer surface in the regenerator, by the addition of fresh make-up stone to the regenerator, or by controlling the rate of lime circulation between the gasifier and regenerator is investigated.

The important operating variables in the regenerative operation are the air/fuel ratio, the gasifier and regenerator temperatures, the limestone particle sizes, the gasifier and regenerator fluidization velocities and bed depths, the stone residence times in both the gasifier and regenerator, and the limestone make-up rate. The effects of these variables are interrelated with the temperature control schemes used and the fuel and limestone properties.

An important phenomenon which takes place in the gasifier during the cracking-partial oxidation is the deposition of carbon, or coking, on the gasifier lime. Carbon deposition affects the sulfur removal efficiency, the thermal efficiency of the system, and the overall operation of the gasifier and regenerator when in the regenerative mode. Minimizing the rate of carbon deposition is, therefore, an important design consideration.

The basic elements of the once-through operation, shown in Figure 10.2, are a gasifier vessel and a sulfate generator, or predisposal vessel. The operation of the once-through gasifier is the same as that of the regenerative operation gasifier, while the sulfate generator operates similarly to the regenerator but at a lower temperature ($\sim$ 1500°F). Thus, the sulfated lime from the gasifier is converted to calcium sulfate rather than calcium oxide. The calcium sulfate may be disposed of while the gas stream from the sulfate generator is sent to the gasifier. The once-through operation requires a limestone addition rate three to four times that used in the regenerative operation.

The same five temperature control methods proposed for the regenerative gasifier are considered for the once-through gasifier, while the sulfate generator temperature is controlled by one of two proposed schemes: heat transfer surface or excess air circulation.

The same operating variables are important in the once-through operation as in the regenerative operation, and carbon deposition is again a major factor in determining the sulfur removal efficiency and the overall system operation. However, in the once-through operation the carbon deposition rate does not affect the system thermal efficiency because the thermal energy of the deposited carbon is recycled as sensible heat from the sulfate generator back to the gasifier. The two conceptual modes of operation described illustrate the basic characteristics of the systems that have been considered.

PROCESS FUNDAMENTALS

Processing Steps

As similarly demonstrated with coal, here the basic fluidized bed residual oil gasification or CAFB process includes four processing steps:

        (1)   Oil gasification/desulfurization;
        (2)   Lime regeneration;
        (3)   Waste stone processing; and
        (4)   Sulfur recovery (optional).

A number of variations of these processing steps have been considered in process evaluations; for example, a once-through limestone mode of operation where the sorbent is highly utilized in the gasifier, up to 90%, and the sorbent regeneration step is eliminated. These concepts have been illustrated in Part 1 of this book (Figure 6.1) and are essentially the same for residual oil as for coal.

The regenerative mode was selected as the demonstration plant basis for commercial assessment of the CAFB process because it had been successfully tested on a pilot-plant scale ($\sim$ 1 MW), and permits operating with the minimum quantity of sorbent. A dry sulfation process which contacts the waste stone with the regenerator sulfur dioxide product-gas and eliminates the sulfur recovery step was selected for the demonstration plant preliminary design and commercial assessments.

A summary of the gasifier conditions, reactions, and performance is shown in Table 10.2. Residual oil (atmospheric or vacuum bottoms) is injected into a bed of lime (870°C/1600°F) along with ½ to 1½ times the stoichiometric amount of make-up limestone and about 17 to 25% of the air required for complete combustion of the oil. The oil cracks, carbon is deposited on the lime particles, the fresh limestone calcines, and hydrogen sulfide (and other sulfur compounds) are released.

TABLE 10.2: SUMMARY OF GASIFIER CONDITIONS AND PERFORMANCE

Conditions

    (1)   Temperature — 870° to 970°C (1600° to 1778°F);
    (2)   Pressure — atmospheric;
    (3)   Air/fuel ratio — 17 to 25% stoichiometric;
    (4)   Fresh limestone rate — 0.5 to 1.5 times stoichiometric;
    (5)   Bed sulfur content — 4 to 6 wt % sulfur as CaS;
    (6)   Bed carbon content — 0.1 to 0.3 wt % carbon;

(continued)

TABLE 10.2: (continued)

(7)   Fluidization velocity — 1.2 to 1.8 m/sec (4 to 6 ft/sec); and
(8)   Bed depth — 0.6 to 1.2 m (2 to 4 ft).

Reactions (two-region fluidized bed)

Oil thermal cracking $\rightarrow$ Carbon + H_2 + hydrocarbons + H_2S, CS_2, COS

Oxidizing Region

$$C + [O_2] \rightarrow CO_2, CO$$
$$CaS + 3/2\ O_2 \rightarrow CaO + SO_2$$
$$CaS + 2O_2 \rightarrow CaSO_4$$
$$Oil + O_2 \rightarrow CO_2, H_2O$$

Reducing Region

$$CaCO_3 \rightarrow CaO + CO_2$$
$$H_2S + CaO \rightarrow CaS + H_2$$
$$CaSO_4 + hydrocarbons \rightarrow CaS$$

Performance
(Demonstrated in Batch Tests and Pilot-Plant Runs)

(1)   90% sulfur removal or greater;
(2)   Removes 100% of fuel vanadium, 75% nickel, 20% sodium;
(3)   Produces fuel gas having high heating value 7.450 kJ/m^3 (200 Btu/scf or greater);
(4)   Combustion of the fuel gas possible with low excess air; acts as two-stage combustor which yields reduction in nitrogen oxides (from 270 ppm to 155 ppm on pilot plant);
(5)   High reliability demonstrated;
(6)   Limestone makeup rates down to 0.5 of stoichiometric are possible, nominally operates at stoichiometric makeup in the regenerative system; and
(7)   Carbon deposition (tars) and lime accumulation may occur in the fuel-gas lines and cyclones.

Source: PB 241 834

The deposited carbon, and some fuel oil, is combusted while the hydrogen sulfide reacts with the lime to form calcium sulfide. A number of other side reactions also occur. The gasifier temperature is controlled by the air/fuel ratio and either stack-gas recycle, water, or steam injection. Control not only of sulfur dioxide but also of nitrogen oxides and fuel oil metals is realized.

The details of the regeneration chemistry are summarized in Table 10.3 along with the regenerator performance. The fundamentals of the regenerator chemistry are illustrated in Figure 10.3. The calcium sulfide produced in the gasifier is circulated to the regenerator and is contacted by air at 1050° to 1100°C (1922° to 2012°F) to generate calcium oxide, some calcium sulfate, and a sulfur dioxide gas of about 8 to 10 volume percent sulfur dioxide. Deposited carbon is also combusted. The regenerated sorbent is recycled to the gasifier.

TABLE 10.3: SUMMARY OF REGENERATOR CONDITIONS AND PERFORMANCE

Conditions

(1) Temperature — 1050° to 1100°C (1922° to 2012°F);
(2) Pressure — atmospheric;
(3) Fluidization velocity — 1.2 to 1.8 m/sec (4 to 6 ft/sec);
(4) Bed depth — 0.61 to 1.22 m (2 to 4 ft);
(5) Bed sulfur content — 0.5 to 1 wt % sulfur as CaS;
(6) Bed carbon content — 0% carbon; and
(7) About 0.1 mol percent oxygen in product gas.

Reactions

$$C + O_2 \rightarrow CO_2$$
$$CaS + 3/2\ O_2 \rightarrow CaO + SO_2$$
$$CaS + 2O_2 \rightarrow CaSO_4$$
$$\tfrac{1}{4}\ CaS + \tfrac{3}{4}\ CaSO_4 \rightarrow CaO + SO_2$$

Performance

(1) SO_2 content of product gas — 8 to 10 mol percent; and
(2) $CaSO_4$ production in regenerator — amount produced depends on stone carbon content, operating temperature, excess oxygen level.

Source: PB 241 834

FIGURE 10.3: REGENERATOR FUNDAMENTALS

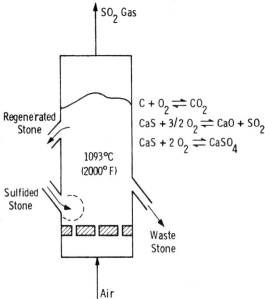

Source: PB 241 835

The rate at which the sorbent is circulated between the gasifier and regenerator controls the regenerator temperature. The sulfur dioxide is at a concentration suitable for sulfur recovery or sulfuric acid production.

A number of processes can be used for converting the spent solids from the regenerator into an environmentally acceptable form for disposal or by-product utilization. The regenerator material is about 92 wt % calcium oxide, about 3 wt % calcium sulfide, about 3 wt % calcium sulfate, and the remainder inert compounds.

Examples of options which might be utilized, depending on market conditions and power plant location, are dry sulfation of the waste with the regenerator sulfur dioxide, dead-burning of the waste solids, and dry oxidation of the calcium sulfide followed by recarbonation of the waste or silica sintering, all of which produce a dry solid product. The options of wet carbonation of the stone or wet sulfation were also considered but are considered less attractive. The waste solids may also be valuable without further processing for their calcium oxide or their vanadium content, which may be about 1 wt % in some cases.

Utility Application

It is indicated that, overall, the once-through operation may be somewhat more attractive to a utility customer than the regenerative operation. Capital investment is reduced with the once-through operation, and, although the limestone feed rate is probably up to three times the rate with regenerative operation, the operating costs for once-through operation may be less than those for the regenerative operation.

The capital investment and fuel adder costs of regenerative and once-through operations can be plotted as a function of the sulfur removal efficiency, the limestone cost, the fuel-oil cost, and the plant capacity factor, and a complete cost breakdown for once-through and regenerative operation calculated. Once-through operation appears to have fewer technical problems and is overall a simpler process than regenerative operation.

A comparison is made in Table 10.4 between atmospheric-pressure oil gasification/desulfurization and the alternative schemes of low-sulfur oil and stack gas cleaning indicates that oil gasification/desulfurization compares favorably with low-sulfur oil and stack gas cleaning, based on cost estimates. When capital costs and operating costs are compared for new and retrofit systems, it is shown that a reduction of about 40% in capital costs involved in stack gas cleaning is estimated for new and retrofit gasification/desulfurization systems.

Operating costs are about the same for once-through stack gas cleaning and regenerative gasification/desulfurization with sulfur recovery. Once-through gasification/desulfurization may reduce the operating costs 30 to 50% as compared to stack gas cleaning. Cost estimates indicate that the operating cost with low-sulfur fuel oil are about 30 to 50% greater than that for gasification/desulfurization. These conclusions are based on the desulfurization of high-sulfur residual oil (3 wt % sulfur) and may be altered if a lower sulfur oil is considered (1 to 1.5 wt % sulfur).

TABLE 10.4: ASSESSMENT OF FLUIDIZED BED OIL GASIFICATION/DESULFURIZATION[a]

	Low-Sulfur Oil	Stack Gas Cleaning	Oil Gasification	
			Regenerative Operation	Once-through Operation
Cost				
Capital, $/kw				
New	---	25-40	12-15	8-10
Retrofit	[b]	40-75	22-27	18-22
Fuel Adder, ¢/10^6 Btu				
New	25-35	11-14	9.5-16.0	9-10.5
Retrofit	25-35[b]	14-20	11-18	10.5-12.5
Efficiency (thermal)	---	0.95-0.98	0.89-0.96[c]	0.96-0.97[c]
Environmental Factors				
SO_2, (lb/10^6 Btu)	0.35	0.45	0.35	0.35
NO_x, (lb NO_2/10^6 Btu)	0.40	0.8	0.16	0.16
Particulates, (lb/10^6 Btu)	0.06	0.02	0.02-0.1[d]	0.02-0.1[d]
Solid Waste, ft^3/MW-day	---	25	15	45
S Removal				
Stone	NA		limestone	limestone
Ca/S	NA		~ 15[e]	3.0
Make-up Ca/S	NA		1.0	NA

[a]Basis: 3% sulfur, 90% sulfur removal, 600 MW capacity, 80% load factor, (1973 $).
[b]Equipment modifications are required when converting from gas or coal to low-sulfur oil.
[c]Overall efficiency is dependent on mode of temperature control.
[d]0.02 figure based on installing electrostatic precipitator (ESP).
 0.1 figure based on installing high efficiency cyclone before burners and no ESP.
[e]Ca/S rate dependent on regenerator temperature control scheme.

Source: PB 233 101

When environmental factors are compared in Table 10.4, the low-sulfur oil is advantageous since capital costs are limited to possible boiler modifications necessary when changing from gas or coal to low-sulfur oil. On the other hand, operating costs are higher than those for oil gasification/desulfurization or stack gas cleaning.

Capital costs are extremely high with stack gas cleaning, especially on the retrofit case, while operating costs are very near those which have been estimated for oil gasification/desulfurization. Some advantages of atmospheric-pressure oil gasification over stack gas wet scrubbers are listed below.

Advantages of Atmospheric-Pressure Oil Gasification Over Stack Gas Wet Scrubbers

(1) Corrosion and fouling problems minimized in SO_2 removal process and boiler (minimum SO_x and V);
(2) No flue gas reheat required;
(3) Uses crushed limestone — no limestone pulverizing system needed;
(4) Simplified disposal — dry solids and no disposal pond;
(5) More compact system;
(6) Reduced structural costs;
(7) Lower auxiliary power requirement;
(8) Reduced energy cost;
(9) Improved NO_x control; and
(10) Potential market for spent CaO.

Industrial Boiler Application

Application of the atmospheric-pressure oil gasification system has also been considered for industrial boiler operation. Capital and operating costs have been projected and compared with alternative sulfur control methods: low-sulfur oil, fluid bed combustion, and stack gas cleaning. Cost and feasibility factors such as load factor, space, reliability, fuel availability, sulfur in fuel, nitrogen oxide, particulates, solids disposal, efficiency, operating cost, and capital cost were also identified.

A study of these factors indicates that industrial boilers will continue to use clean fuels as long as they are available. If clean fuels are not available, fluidized bed oil gasification or combustion offers economical advantages over competitive systems (e.g., stack gas cleaning).

The fluidized bed systems are also able to utilize low-grade fuels (e.g., high-sulfur vacuum residuum) which cannot be easily handled in conventional systems. Thus, industrial applications of fluidized bed oil gasification for industrial fuel gas and/or steam generation offer economical and environmental advantages and should be considered if clean fuels are not available.

ASSESSMENT

Summary

The atmospheric-pressure fluidized bed residual oil gasification/desulfurization process (CAFB) produces a clean, low-heating-value fuel gas for firing in a conventional boiler. The integrated process has been operated successfully in a 750 kW pilot plant unit, and the process has demonstrated the ability to meet environmental emission standards for sulfur oxides, nitrogen oxides, and particulates.

The work described was directed toward the completion of a preliminary design and cost estimate for a 50 MW demonstration plant and a 200 MW plant design and cost estimate; several process and design options have also been evaluated.

Work has also been carried out to develop process design data and criteria. Experimental tests have been performed to identify candidate limestone sorbents (14 limestones tested); to obtain data on sulfur removal, sorbent regeneration, and spent stone processing (five process concepts tested); and to assess the environmental impact of the spent sorbent.

Alternative systems to the CAFB process are stack-gas desulfurization via wet scrubbing and hydrodesulfurization (HDS) of residual fuel oil, including vacuum residuals. Unfortunately, comprehensive cost and performance data on these systems are limited.

Conclusions

The primary conclusions are:

(1)   There will be sufficient quantities of high-sulfur residual fuel oils, and there are existing power plants such that a market exists for retrofitting electrical utility power plants to utilize high-sulfur fuel oils;

(2)   The process offers the potential to utilize low-grade petroleum or synthetic fuel fractions with high metals content which cannot be economically utilized in alternative processes for producing clean power with conventional boilers;

(3)   On the basis of the development work which has been carried out, a 50 MW plant can be built;

(4)   The economics of the CAFB process are sufficiently attractive that the process merits development for utilizing low-grade, high metals content, residual fuel oils; and

(5)   A pressurized fluidized bed residual oil gasification process for firing a combined cycle power plant is lower in cost than an atmospheric-pressure retrofit system. The total power plant cost for a pressurized power plant system will be significantly lower than for a conventional power plant.

The following conclusions are based on assessments of the market, technology, process economics, and the demonstration plant design and cost estimate.

Market:

(1)   Atmospheric-pressure operation of the CAFB process

is most applicable to the electrical utility industry as a boiler retrofit for oil- or gas-fired boilers.
The atmospheric-pressure process is not generally
attractive for new boilers or retrofit of coal-fired
boilers because of the trend toward coal. The
largest market is for boiler sizes ranging from 50
to 400 MW;

(2) Vacuum bottoms or other low-grade high metals
content fuels are the most likely to be available
for the CAFB process;

(3) Limestone availability for CAFB may be somewhat
more restricted than for slurry scrubbing processes
due to more stringent requirements on sorbent
physical properties with CAFB; and

(4) Potential markets for CAFB by-product/waste stone
are not known.

Technology:

(1) Environmentally, CAFB appears superior to lime
and limestone slurry flue-gas scrubbing because of
the reduced impact of nitrogen oxide, solid waste,
and process water requirements;

(2) Higher fuel efficiencies are realized with CAFB
than with HDS;

(3) The ability of CAFB to utilize low-grade petroleum
or synthetic fuel fractions with high metals content
provides the potential for a fuel source which may
not be feasible with HDS or stack-gas cleaning
processes;

(4) CAFB has the potential to offer high reliability and
lower space requirements than lime/limestone
slurry scrubbing;

(5) It has to be established whether there is, in general,
sufficient physical space at a majority of boiler
plant sites for either CAFB or slurry scrubbing
retrofit; and

(6) The utilization of a clean fuel is the most convenient technological option available to the utility.

Economics:

(1) The capital cost of CAFB is 25 to 50% greater than
that of limestone scrubbing and is comparable to
regenerative stack-gas cleaning costs;

(2) The fuel adder for CAFB is competitive with limestone scrubbing if low-grade fuels available to
CAFB are 10 to 20¢/10^6 Btu cheaper than fuels
suitable for firing in a plant with limestone scrubbing; and

(3) HDS of vacuum bottoms with low metals content is
not competitive, even though operating at a higher
load factor, unless a unit larger than 25,000 bbl/d

is built to supply more than three 200 MW boilers, or a unit larger than 35,000 bbl/d is built to supply two 500 MW boilers. Thus, HDS requires more immediate commitment of capital than would a single CAFB unit.

Demonstration Plant Design and Cost:

(1) The design basis for the gasifier-regenerator system is based on the data obtained in a 750 kW pilot plant unit;

(2) Experimental tests were carried out to evaluate five spent stone processing options. These data indicate:

(a) Sulfur dioxide and spent limestone from the regenerator can be combined in a dry sulfation process to form calcium sulfate which is suitable for disposal. More than 90% sulfation was obtained in laboratory tests;

(b) High-temperature (1400°C) sintering of the spent sorbent from the regenerator is a feasible method for processing the spent limestone for disposal;

(c) Slurry recarbonation of spent limestone flue gas will produce a material which is environmentally suitable for disposal and may be marketable;

(d) Treatment of the spent limestone with dilute sulfuric acid will produce material containing primarily calcium sulfate (94% sulfation) which is environmentally suitable for disposal; and

(e) Oxidation of the calcium sulfide to calcium sulfate followed by high-temperature recarbonation of the calcium oxide is not practical for a regenerative process and is feasible in a once-through sorbent process only if the residual calcium sulfide is not released.

(3) The concept chosen for a CAFB demonstration plant used dry sulfation of the waste sorbent to eliminate the sulfur recovery step. This appears to be the most economical processing concept for CAFB other than once-through operation with direct disposal of the utilized sorbent;

(4) Screening tests were carried out on 14 candidate limestones for the demonstration plant; limestone 1359 and aragonite were selected as the candidate sulfur sorbents;

(5) High excess capacity factors were applied to the demonstration plant design to assure flexibility of operation;

(6) The most important cost component in the plant is the hot fuel-gas piping system; other costly components are the gasifier cyclones and the fines recycle system; and

(7)   Because of cost considerations the four fuel-gas burn-
       ers should be placed in a single boiler face rather
       than tangentially in the four corners.

It is therefore recommended that development of the atmospheric-pressure fluidized bed residual oil gasification/desulfurization process be directed toward the
utilization of low-grade, high-metals-content fuels, such as vacuum bottoms,
which are not compatible with conventional boiler/stack-gas cleaning systems
and that fuel availability for the process be investigated to assess market need
and market potential for the process. Fuels to be investigated should include
but not be limited to vacuum bottoms, Venezuelan bitumens, residual oil from
shale, tar sands, and liquefaction of coal.

SEWAGE, INDUSTRIAL AND
OTHER
WASTE TREATMENT PROCESSES

The material in this chapter was excerpted from:

Chemical Engineering Progress
CONF-751213-3
ERDA-76-69
ICP-1088
PB 206 892
PB 211 323
RFP-2016
RFP-2471

For a complete bibliography see p 263.

SEWAGE SLUDGE INCINERATION

Evaluation of Sludge Incineration

More and more, as action is taken to solve environmental problems, it is discovered that the solutions chosen have produced new and different problems. Such is the case concerning the ultimate disposal of sludge from sewage treatment plants.

Primary sewage treatment is designed to remove the bulk of the solids present in the sewage. The removed solids, in the form of sludge, are either disposed of directly from the primary process or receive their final disposal after combination with sludges from secondary treatment processes.

In certain areas, particularly on the East Coast, the practice has been to barge sewage sludge to sea, for ultimate disposal into the ocean depths. It is this form of disposal which has been the subject of official concern and action because of its apparent adverse affect upon life on the ocean floor.

In October 1970, the Council on Environmental Quality recommended, in its

"Ocean Dumping - A National Policy," that ocean dumping of sludge should be phased out as an ultimate disposal practice. The State of New Jersey has also taken strong legal steps to curtail sludge dumping at sea. If unrestricted disposal at sea does not continue to be an acceptable practice for sludge removed in sewage treatment plant operations other alternatives must be considered. The Environmental Protection Agency has established a task force to evaluate sludge incineration as an acceptable alternative to sea disposal.

With incineration of sludge, of course, there is an inherent potential for air pollution and it is this potential which has dominated the evaluations.

An EPA Task Force study program investigated fluidized bed and multiple hearth incinerators across the country. With respect to process flows, the most significant difference is that the major part of the ash leaves from the bottom of the multiple hearth furnace, whereas in the fluidized bed incinerator, all the ash is carried overhead and is removed from the bottom of the scrubber. All of the incinerators incorporated scrubbing devices for control of particulate matter. As a result of this initial examination, three incinerators were chosen for detailed sampling and analysis.

One of these installations, located in Barstow, California, utilized the fluidized bed principle. Its input included only raw sludge from primary treatment that came from essentially domestic (residential) loads. The other two were multiple hearth incineration systems and are discussed here only for comparison purposes.

Sampling Results

A summary of the information which has resulted from the sampling program is as follows.

Stack Gas Analyses for Particulates, NO_x, SO_2 and Visible Emissions: Analyses of the stack gases indicated that the overall dry filterable particulate average concentration did not exceed 0.07 grain per standard cubic foot. Oxides of nitrogen emissions were obtained; the higher overall test average emission concentration did not exceed 175 ppm.

Sulfur oxides measured concentrations did not exceed 13 ppm as an average emission during the test period. Visible emissions were less than 10% opacity. These tests results indicate that a well designed and operated sludge incinerator is able to achieve acceptably low emission concentrations for the common pollutants.

Stack Gas Analyses for Major Metals: Measurable concentrations of metals were found in the stack gas material. Lead and other elements such as cadmium, chromium, copper, nickel and vanadium were measured in the particulate matter analyses.

Input Sludge Analyses: As would be expected from the stack gas samples, an examination of the input sludge showed measurable quantities of the heavy metals and other important pollutant compounds. A summary of the major and minor elements in sludge is given in Table 11.1. The elements cadmium, beryllium, nickel, copper, lead, chromium, vanadium and mercury are toxic and accumulate in the human body; all, with exception of beryllium and nickel, were found in the input sludge analyses.

TABLE 11.1: SUMMARY OF MAJOR AND MINOR ELEMENTS IN SLUDGE (MILLIGRAM PER GRAM DRIED SLUDGE)

Elemental Analysis	Primary Sludge			Activated Sludge			Digested Sludge		
	Average	Range	No. in Sample	Average	Range	No. in Sample	Average	Range	No. in Sample
Al	5.10	10.78-1.83	3	10.0	17.0-4.35	3	17.86	36-7.75	4
Sb	<1.24	<1.49-<0.83	–	–	–	–	0.897	0.984-0.81	2
As	2.25	5.0-0.11	11	1.20	2.22-0.101	3	–	–	–
Ba	0.0025	0.0030-0.0017	3	1.15	3.0-0.22	4	1.36	4.01-0.10	15
Be	0.104	0.15-0.07	11	0.0035	0.0044-0.0026	2	0.0025	0.0065-0.0012	11
B	<0.188	<0.30-0.0034	4	0.070	0.22-0.006	9	0.046	0.149-0.003	17
Cd	–	–	–	0.35	0.44-0.26	2	0.264	0.50-0.001	10
Ca	2.05	9.0-0.08	15	13.0	18.0-9.0	7	33.5	112.0-4.9	15
Cr	0.217	0.5-0.05	6	4.31	17.0-0.1	8	2.28	11.0-0.10	28
Co	2.00	6.0-0.0083	17	0.0016	0.0016	1	–	–	–
Cu	0.063	0.1-0.01	7	1.10	2.6-0.372	13	1.65	16.0-0.10	39
Ga	16.1	20.0-2.86	12	0.05	0.05	1	0.05	0.05-0.05	3
Fe	1.01	2.14-0.33	3	40.5	96.6-4.83	9	30.65	60.6-10.09	19
Pb	10.6	15.0-5.0	8	1.52	2.09-0.51	3	1.89	7.52-0.18	18
Mg	0.781	1.0-0.16	11	7.04	10.9-3.01	7	7.49	13.0-1.0	17
Mn	0.0046	0.006-0.0030	2	0.310	0.93-0.065	9	0.976	6.04-0.06	17
Hg	0.362	1.0-0.05	11	0.016	0.020-0.012	2	–	–	–
Mo	0.522	2.0-0.0014	17	0.197	0.89-0.006	8	0.254	1.29-0.002	24
Ni	3.78	6.83-1.49	3	0.378	2.0-0.04	8	0.372	3.0-0.03	27
P	–	–	–	19.9	32.2-11.07	8	12.75	25.13-1.18	15
K	0.243	1.0-0.08	11	4.21	7.16-2.49	6	2.76	6.15-0.83	13
Si	3.96	10.0-0.5	8	39.5	39.5	1	162.0	334.0-73.0	3
Ag	0.13	0.14-0.12	3	0.150	0.22-0.1	3	0.195	0.50-0.08	4
Na	–	–	–	4.44	7.88-1.0	2	6.15	10.0-2.0	5
Sr	0.95	2.0-0.5	8	0.155	0.21-0.10	2	0.26	0.26	1
S	14.8	20.0-5.0	8	10.1	11.6-7.6	6	12.3	32.5-1.64	14
Sn	2.09	15.0-0.3	11	0.5	0.05	1	0.60	0.70-0.50	3
Ti	6.87	35.0-0.34	18	11.8	20.0-0.50	3	14.2	20.0-1.0	3
V	1.72	10.9-0.3	8	0.70	0.89-0.51	3	5.20	10.0-0.32	4
Zn	–	–	–	3.29	6.3-0.13	13	4.04	11.0-0.5	39
Zr	–	–	–	10.0	10.0	1	2.03	5.0-0.10	3

Source: PB 211 323

In addition, analyses were made for major pesticides in the input sludge. At Barstow, measurable concentrations of PCB, dieldrin, chlordane, and DDD were found.

Analyses of Incinerator Ash: Metal analyses were also made of the ash remaining after incineration of the sewage sludge. The data showed that all of the mercury in the sludge is apparently lost in the stack gases, and that lead compounds are apparently carried off in the fly ash. A comparison of sludge ash composition with a coal fly ash showed the sludge ash to have concentrations of copper, zinc, chromium and lead one or two orders of magnitude higher than coal fly ash. No pesticides or PCB materials were found in the ash or scrubber water.

It was therefore concluded that ash resulting from the incineration of sewage sludge would constitute no significant threat to the environment if ultimate disposal of the ash was by acceptable sanitary landfill practice as outlined by "Sanitary Landfill Design and Operation," a United States Environmental Protection Agency report.

Because of the importance which was given to the measured presence of various metals found, sludge samples were taken at additional sewage treatment plants, and analyzed for metals content, and these plants included units with significant industrial inputs.

Again, tests at the additional plants indicated the presence of biologically non-degradable metals.

Incinerator Testing Programs

The Office of Air Programs undertook a survey and test program for sewage sludge incinerators in order to obtain information both for the setting of source performance standards and, also, for the purpose of the EPA Task Force on Sewage Sludge Incineration to determine if this is a viable disposal method as far as its total environmental impact is concerned. The following material provides basic information with regard to this program.

To satisfy these two requirements an approach to establish a test program was planned in two phases. Phase I involved inspection visits to sixteen existing incinerators that were reported to be capable of consistently meeting the most stringent particulate emission limitations (0.1 grain/scf). Based on this inspection survey, three plants which incinerate primarily residential sludge were selected for testing to provide preliminary data to the Environmental Protection Agency Sludge Incineration Task Force.

At each of the sixteen sites, samples were collected of the feed sludge, scrubber water inlet and outlet, and ash residue. The samples were sent to the Taft Water Research Center for analyses with particular emphasis on heavy metals.

Phase II was to consist of a minimum of three additional tests on plants selected after a more thorough review of existing incinerators and examinations of the analytical results of the Phase I study. An additional fourteen incinerator sites were visited during Phase II. Only one of these plants, the N.W. Bergen plant at Waldwick, New Jersey, was deemed to be adequately controlled and suitable for testing. The State of New Jersey had tested this plant using the EPA test method. Particulate emissions measured during the State agency test ranged from 0.02 to 0.048 grains per cubic foot (total catch).

Source tests were completed on the incinerators selected for study. Discussed here are the two plants selected for study which utilized fluidized bed incinerators. They are: Barstow, California which has a fluo-solid (fluidized bed) reactor; design capacity of 500 lb/hr dry solids; and operated at 95% capacity during test. The control device is an ARCO single cross flow perforated plate type impinjet scrubber. There is a 4.0" H_2O pressure drop.

The other is in northwest Bergen County, Waldwick, New Jersey, which also has a fluo-solid (fluidized bed) reactor; design capacity of 1,100 lb/hr dry solids; and operated at 100% capacity during test. The control device is a Peabody Five Plate impinjet scrubber. There is a 20.0" H_2O pressure drop. Samples of the sewage sludge residue (ash), and stack gas were collected and chemically analyzed with particular emphasis on heavy metals.

The air samples were collected and analyzed for NO_x and SO_2 concentrations and particulate grain loading. The results are reflected in the Test Data Summary, shown in Tables 11.2 and 11.3.

TABLE 11.2: SUMMARY OF RESULTS—BARSTOW, CALIFORNIA PLANT

	. . . . Run Number		
	1	2	3
Stack flow rate, scfm	1,190	1,170	1,230
Percent CO_2, vol. % dry	8.8	9.9	9.1
Percent excess air at test point	38.7	50.7	59.4
SO_2 emission, ppm	10.71	14.10	13.16
SO_2 emission, lb/hr	0.1274	0.1650	0.1619
NO_x emission, ppm	161.64	42.18	173.20
HCl emission, ppm	1.64	2.78	2.15
Particulate emission, filter			
Gr/cf at stack conditions	0.0468	0.0650	0.0464
Gr/scfd	0.0551	0.0766	0.0545
Pounds per hour	0.562	0.768	0.574
	2.2 lb/ton	3.24 lb/ton	2.84 lb/ton
Particulate emission, total			
Gr/cf at stack conditions	0.0565	0.0729	0.0556
Gr/scfd	0.0665	0.0859	0.0653
Pounds per hour	0.678	0.861	0.688
Process conditions			
Sludge feed to furnace,			
lb/hr dry solids	510.0	474.0	403.0

Source: PB 211 323

These sewage treatment plants that have been source tested have been plants that incinerate municipal or residential type sludge. To obtain data to aid the EPA Task Force in evaluating the environmental impact of sludge incinerators that burn a high percentage of industrial waste, sludge samples were collected from a number of plants that treat industrial waste and analyzed to determine composition and heavy metal content.

TABLE 11.3: SUMMARY OF RESULTS—WALDWICK, NEW JERSEY PLANT

	. . . . Run Number		
	1	2	3
Stack flow rate, scfm	3,650	3,788	3,513
Percent CO_2, vol. % dry	4.0	5.1	4.0
Percent excess air at test point	–	–	–
Particulate emissions, total			
Gr/cf at stack conditions	–	–	–
Gr/scfd	0.020	0.031	0.048
Pounds per hour	0.61	1.0	1.45
Process conditions			
Sludge feed to furnace			
lb/hr, dry solids	650	650	650

Source: PB 211 323

Preliminary analyses of industrial type sludge indicate that the heavy metals found in industrial waste are basically the same as in residential waste, varying only in concentration. Using the sludge feed and stack emission analytical data collected from tests of residential type sludge incinerators, heavy metal concentrations in the stack gas of industrial sludge incinerators can be calculated when the concentration of a particular heavy metal in the sludge feed is known.

SYSTEMS FOR MUNICIPAL SEWAGE

Dorr-Oliver, Inc. has been active in the development and marketing of fluid bed systems during the past 30 years, for such unit operations as roasting, calcining, drying, sizing, and incinerating.

For the incineration of various sludges and other waste material, the company calls its systems FluoSolids Reactor Systems. Their first fluid bed incinerator was a small, 4 foot diameter unit for handling 210 lb of primary sewage sludge solids. From this beginning, the basics for the incineration of municipal sewage sludge were developed.

Cold Windbox System

The cold windbox flowsheet is illustrated in Figure 11.1. Cold air is supplied directly to the reactor windbox and dewatered sludge solids or fluids are introduced directly into the fluid bed by pumps or screw feeders. Auxiliary fuel, oil or gas as required, introduced into the silica sand fluid bed, is combusted along with the organic components of the sewage sludge to provide complete destruction of the waste solids. The sludge water as steam, the gaseous products of combustion of the organics and fuel, and the fine inert ash solids suspended in these gaseous products exist from the reactor to a gas scrubber. These gases are cleaned and cooled and exit to atmosphere at 120°F with less than 0.05 grain per dry standard cubic foot of particulate solids.

FIGURE 11.1: FLUOSOLIDS DISPOSAL SYSTEM USING A COLD WINDBOX

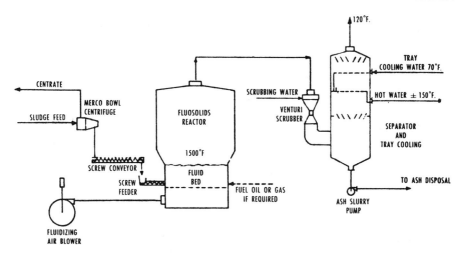

Source: *Chemical Engineering Progress,* October 1976.

Hot Windbox System

The hot windbox flowsheet is shown in Figure 11.2. Flow is much like that in the cold windbox except air temperature is raised first in a heat exchanger to 800° to 1200°F by the incinerator off-gases, which are thereby cooled to ±1000°F. With air preheating, the system has an improved thermal efficiency.

The Btus from the incinerator exit gases supply heat to the incinerator for evaporation of sludge water without the consumption of fuel or of oxygen from the fluidizing air.

The heat required to evaporate water with this design is 2,500 to 2,600 Btu per pound of water, compared to 3,700 to 4,100 Btu per pound evaporated for the cold windbox system.

Waste Heat Boiler System

The third system, shown in Figure 11.3, does not preheat the fluidizing air. Instead the sensible heat in the reactor exhaust gases is extracted in a waste heat boiler to generate steam. Incinerator gases are cooled to about 650°F in the waste heat boiler, move to a gas scrubber for final cleaning and cooling, and then are removed by an exhaust fan. A hot Cottrell precipitator or an electrofilter could also be used for final cleaning of the incinerator exit gases.

FIGURE 11.2: THE FS SYSTEM USING A HOT WINDBOX

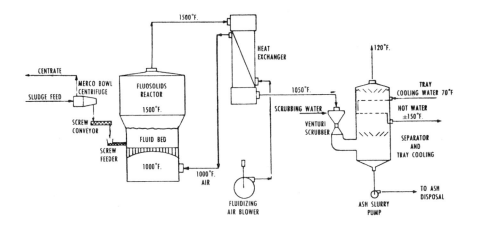

FIGURE 11.3: THE FS SYSTEM WITH WASTE HEAT BOILER

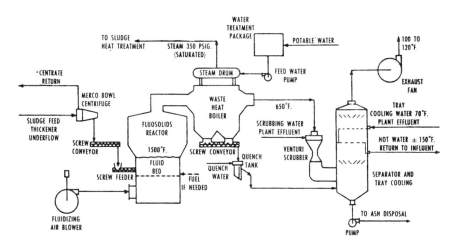

Source: *Chemical Engineering Progress,* October 1976.

Steam used for sewage sludge heat treatment is generated at about 350 lb/in^2 gauge and saturated at 437°F. Any excess steam may be used for heating or power generation or for driving rotary equipment in the incinerator plant.

Municipal sewage sludges being incinerated in these systems include: primary sludge, secondary or activated sludge, digested sludge, heat-treated sludge, sludge containing precipitated chemicals resulting from phosphorus removal from the sewage, and various combinations of these sludges.

Plant capacities per incinerator unit vary from as little as 210 to a maximum of 3,500 lb/hr of total dry solids. Total dry solids in the sludge feed for the incinerator range from a low of 10% to a high of 40%. Reactor sizes (inside diameter of the refractory in the freeboard) range from 4 to 21 feet.

Neutral Sulfide Semichemical Waste Liquor

The spent pulping liquor from neutral sulfite semichemical (NSSC) paper mills represents a waste stream which formerly presented a difficult disposal problem. By using the FluoSolids incineration system, spent NSSC liquor, a mixture of digested organics representing about half the weight of the original wood together with the salt residue of the cooking liquor, can be incinerated. The organics are completely oxidized, leaving a salt mixture of Na_2SO_4 and Na_2CO_3 which is in the form of small nodules or pellets.

Sulfur is added to bring the sulfate content up to 90 to 95%, forming a salt cake. The salt cake is sold to another type (kraft) pulp mill as their cooking liquor make-up. This is a good example where the ash product, in the form of nondusty free-flowing pellets, represents a saleable by-product from the incineration of a waste stream. Figure 11.4 details the principal features of this fluid bed incinerator for these waste pulping liquors. Concentrated waste liquor is fed directly into the upper combustion fluid bed where it is oxidized to CO_2 and H_2O, and the ash product, Na_2SO_4-Na_2CO_3, is formed into pellets.

FIGURE 11.4: THE INCINERATOR FOR NEUTRAL SULFIDE SEMICHEM— ICAL WASTE LIQUOR

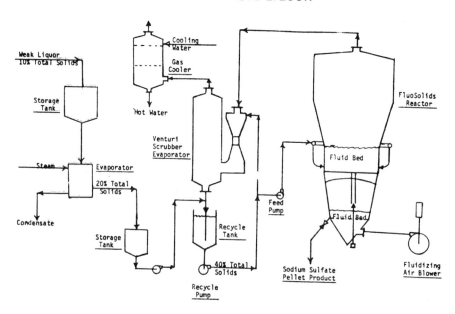

Source: *Chemical Engineering Progress,* October 1976.

A lower fluid bed cooling stage cools the ash product pellets and provides some air preheating. The hot exhaust gases from the combustion compartments, at about 1350°F go to a venturi scrubber-evaporator where they are cleaned and cooled and used to evaporate some of the water from the incoming waste liquor.

Plant capacities for installations of this type range from 6,000 to 12,500 lb/hr of total dry waste liquor solids at a feed concentration to the incinerator of about 40% total dry solids. This corresponds to a water load of 9,000 to 18,750 pounds per hour in the incinerator. The reactors have a 14 to 20 foot inside diameter.

COMBUSTION OF ANTHRACITE WASTES

Combustibility of Anthracite Wastes

The waste material resulting from the mining and cleaning of anthracite coal, principally in northeastern Pennsylvania, contains a substantial amount of residual fuel value which can be recovered in fluidized bed combustors. In the mining region, this refuse material is piled in a densely populated, geographically small area of about 480 square miles.

It is estimated by the U.S. Bureau of Mines (1) that 800 banks containing 910 million cubic yards of refuse can be found within the anthracite mining fields. This material has been deposited from mining operations or from the reject streams of preparation and cleaning plants (called breakers). Many of the banks contain refuse spanning the history of mining in the region. The deepest layers may have been deposited 100 years ago when preparation methods were crude, while surface layers have resulted from much better heavy media washing procedures and are essentially depleted in coal. Some of the older banks have been reworked using the better cleaning methods to recover coal; consequently the characteristics of the refuse varies widely from bank to bank or even within a given bank.

The Morgantown Energy Research Center (MERC) of the U.S. Energy Research and Development Administration has begun a program of examining the combustibility characteristics of these refuse materials in a fluidized bed combustor. The value of this refuse as a fuel is important in the anthracite region because of the decline in mining in the area which has led to a growing dependence on oil.

The refuse banks are aesthetic eyesores which have supplanted valuable land from other uses along with posing potential health and safety hazards from spontaneous combustion. The burning of this material in fluidized bed combustors would provide fuel in a region which needs fuel along with reducing some of the problems presented by the refuse banks.

Experimental Facility and Program

The Morgantown atmospheric pressure fluidized bed combustor is basically a refractory-lined cylindrical combustor of 18-inch internal diameter in the bed region with an expanded freeboard cross section of 24-inch diameter. The combustor is equipped with a horizontal, water-cooled heat exchanger submerged in the bed and a separate water-cooled tube bundle in the freeboard to reduce exit gas temperatures. For temperature control with low heating value refuse fuels, exchange is accomplished by six hairpin loops of one-fourth inch type 310 stainless steel pipe with individual water flow controls.

During operations some tubes may be uncooled, depending upon the heat removal required to maintain bed temperature; thus the alloy pipe is necessary to withstand full bed temperature. Fuel is pneumatically injected into the base of the combustor with room temperature air. Fluidizing air is provided through a plenum which feeds a number of orifices in the conical distributor. The reject solids which are separated from the exit flue gas by the primary cyclone can be reinjected into the bed with an air injector for carbon burnup. Flue gases are further cleaned by the secondary cyclone and parallel bag filters before exiting through the stack. Gases are sampled for on-line analysis at the exit of the combustor. Excess spent bed is withdrawn through the apex of the inverted, conical air distributor with a screw feeder.

An operating period with the combustor is typically of five days, 24 hour per day duration. Startup begins by preheating the empty combustor vessel with a premixed natural gas/air flame through the air distributor.

When operating temperature is reached, the fluidized bed is built by feeding either a 50-50 mixture (by weight) of anthracite coal and inert material (as limestone) or with the refuse directly. When the planned bed depth has been achieved, the natural gas flow is curtailed and the temperature of the bed stabilized by adjusting water flow in the submerged heat exchanger followed by reinjection of the primary cyclone ash. The complete startup procedure requires 2 to 4 hours from cold lightoff to stabilization of temperatures with normal feeding of refuse and reinjected ash.

Two anthracite refuse materials with widely different characteristics have been burned. These include a fine silt refuse from the Powderly Bank, south of Scranton, Pennsylvania, and a much lower quality material from a reworked bank north of Scranton. The silt is the fine reject from a washing operation which contains coal along with finely divided earth minerals.

The second material is the heavy reject from the breaker. It originally consisted of lumps up to four-inch size which had been crushed to minus three-fourth inch for use in backfilling mines to control subsidence. The minus three-fourth inch size was further processed at MERC through a hammer mill to pass a one-fourth inch mesh size.

The fine silt was burned in the size as received. Each of these culms was obtained from outside storage piles and was air dried before being burned. The proximate, ultimate and size analyses of these two materials are shown in Table 11.4.

These refuse materials are two extremes of fuel quality. The silt approaches anthracite coal in analysis and heating value, while the reworked refuse bank is much poorer in quality containing mostly slate and bone (a laminated coal/slate agglomerate of medium carbon value). Similar low grade, coal-related waste materials have been burned in fluidized beds in England in the form of slurries usually with a supplementary fuel required (2)(3).

TABLE 11.4: ANALYSES OF ANTHRACITE REFUSE MATERIALS FOR
FLUIDIZED BED COMBUSTION TEST

	Powderly Bank Silt		Mid-Valley Bank Crushed, Reworked Culm	
Proximate				
Moisture, wt. %	2.8		1.1	
Ash	24.3		63.5	
Volatile matter	6.9		10.4	
Fixed carbon	66.0		25.0	
Ultimate				
Sulfur	0.7		0.5	
Hydrogen	2.7		1.9	
Total carbon	66.1		25.3	
Nitrogen	0.8		0.8	
Oxygen	2.6		6.9	
Heating value, Btu/lb				
(as received	10,380		3,990	
	Direct	Cumulative	Direct	Cumulative
	. . . . .weight percent			
Size, Tyler mesh				
+4	–	–	10.0	10.0
–4+16	2.9	2.9	50.3	60.3
–16+28	12.5	15.4	25.8	86.1
–28+50	24.0	39.4		
–50+100	28.4	67.8	13.9	100.0
–100+200	19.2	87.0	–	–
–200	13.0	100.0	–	–

Source: CONF-751213-3

The fluidized bed combustor is well suited to burning such materials because of
the inherent low carbon in the bed coupled with the good mixing and long
residence time to burn the relatively unreactive carbon found in such waste
materials.

Operation of System

The combustion tests with these two refuse materials demonstrate satisfactory
carbon burnout and smooth sustained operations. Each has its own unique operating requirements to achieve these results, however.

Although the silt contains high carbon and good overall heating value, the fine-
ness of the particles presented problems in retaining the bed material—even at
the low superficial velocity of 2 ft/sec. (Operating conditions are summarized
in Table 11.5.) In the silt runs a deeper bed was required (up to 40 inches
above the highest point on the conical air distributor), along with the presence
of larger inert particles to help retain the fines.

Limestone in a one-fourth-inch to zero size range was used as the larger particles
(not specifically for sulfur capture purposes). Any inert material would have
served the same purpose equally well. Primary cyclone reject ash was recycled
to achieve the carbon burnout. Limestone was added throughout the run to
maintain bed level since most of the ash was carried out of the bed and collected in the cyclones and bag filters.

TABLE 11.5: SUMMARY OF OPERATING CONDITIONS FOR POWDERLY
SILT AND MID-VALLEY ANTHRACITE REFUSE COMBUSTION TESTS

Operating Parameter	Powderly Silt with Limestone	Mid-Valley Refuse without Limestone
Run duration, hr	106	69
Bed temperature, °F*	1,600	1,645
Superficial velocity, ft/sec	2.0	2.9
Fuel feed rate, lb/hr	42.4	117.0
Heat input, Btu/hr-ft^2	250,800	265,200
Air/fuel, scf/lb	77.1	38.6
Limestone rate, lb/hr	6.7	–
Ca/S mol ratio	6.9**	–
Sulfur retention, wt %	78.6	34.0
SO$_2$ emission, lb/10^6 Btu***	0.03	1.65
Carbon utilization, pct	83.4	79.0
Overall heat transfer coeff., hr-ft^2-°F†	–	63.5

   *Bed depth (expanded) was 36 to 40 inches above the highest
     part of air distributor in all runs.
  **Limestone addition in this run controlled by bed height and
     bed temperature, not sulfur capture.
 ***Calculated from retained sulfur in solids.
   †Heat transfer coefficient to submerged water-cooled one-fourth
     inch pipe.

Source: CONF-751213-3

The low sulfur content of the silt, with its high heating value would meet sulfur
emission limits when inherent sulfur retention in the ash is considered (the complete transport of the sulfur into the flue gas would yield 1.34 pounds of SO$_2$
emission per million Btu heat input; retention of only 0.07 pound of sulfur per
million Btu would control emissions at the 1.2 lb SO$_2$/10^6 Btu limit). Using
the limestone, the 79% sulfur retention achieved in the solids limited the emission of SO$_2$ to 0.03 lb/10^6 Btu (see Tables 11.4 and 11.5).

While carryover of the fine silt particles was a problem, the deeper bed and recycle of primary cyclone ash allowed carbon utilization of 83% to be achieved.

The silt combustion test demonstrated that this fine material can be handled at
low velocities in the fluidized bed combustor. The variability of these refuse
piles also became evident with the procurement of this feed material. Whereas
a preliminary grab sample of the Powderly silt indicated that 52% carbon and
39% ash (8,400 Btu/lb) would be contained, the bulk sample (5 ton load), obtained in the same vicinity as the grab sample, proved to be much better quality
(Table 11.4).

The burning of the Mid-Valley bank refuse was a much more severe test of the
combustibility limits for these low grade fuels. In the first test burning this
material, limestone addition was avoided to demonstrate that this high ash material would build and maintain its own bed. Based on the results from this first
run, however, it was evident that some limestone would be required to meet
sulfur emission limits.

In the first run with the Mid-Valley Bank refuse, a marked difference was evident compared with the silt combustion. Whereas ash from the silt was carried out of the bed with the flue gas; the residue remained in the bed with the coarser material. Continuous bed removal at a rate of 40 to 60 lb/hr was necessary to maintain the bed level in the combustor. The difference is evident in comparing the material balances of Tables 11.6 and 11.7—about 60% of the ash is retained in the bed with the Mid-Valley refuse compared with only 20% of the silt ash.

TABLE 11.6: MATERIAL BALANCE—POWDERLY BANK SILT COMBUSTION TEST

Material	Total	Ash	Carbon	Sulfur	Oxygen	Nitrogen	Balance
	. . . . Pounds. . . . .
Refuse	100.0	24.3	66.1	0.7	2.6	0.8	5.5
Air	551.0	0.0	0.0	0.0	128.4	422.6	0.0
Limestone	15.8	9.9	1.6	0.0	4.3	0.0	0.0
Total in	666.8	34.2	67.7	0.7	135.3	423.4	5.5
Bed	9.3	9.1	0.2	0.07	~0	~0	(0.07)
Primary cyclone	2.0*	1.1	0.9	0.02	0	0	(0.02)
Secondary cyclone	25.8	17.1	8.0	0.25	0	0	0.45
Bag filters	10.3	8.1	1.8	0.21	0	0	0.19
Flue gas and loss	619.4	(1.2)	56.8	0.15**	135.3	423.4	4.95
Total out	666.8	34.2	67.7	0.7	135.3	423.4	5.5

*Primary cyclone ash was recycled through most of the period.
**Equivalent SO_2 emission of 0.03 lb per million Btu heat input.

Note: Basis—100 lb refuse in.

Source: CONF-751213-3

Best operation was at higher bed temperature; sustained autothermal combustion was difficult at temperatures below 1600°F. At 1650°F, however, combustion was stable with ash recycle and heat removal in two of the in-bed heat transfer loops. The overall heat transfer coefficient with the exchanger was 63 Btu/hr-ft²-°F (Table 11.5), which is within the range previously reported in the MERC Combustor with a limestone bed at 3 ft/sec superficial gas velocity.

The combustion of both types of refuse was principally within the dense phase fluidized bed. Temperature profiles indicate highest temperatures in the expanded bed with much lower freeboard gas temperature.

The first run with the crushed refuse indicated that combustion of this grade material is possible with the draw off of ash from the bed. The overall carbon combustion efficiency was lower than with the silt (79% compared to 83%). The bed material contained 2 to 3% carbon which appeared to be contained within the slate-like particles where carbon could be seen when the particles were fractured. Calculation of retained sulfur in the solids indicates that some limestone is needed with this refuse despite its low sulfur content (Table 11.5).

TABLE 11.7: MATERIAL BALANCE—MID-VALLEY BANK REFUSE COM-
BUSTION TEST (WITHOUT LIMESTONE)

Material	Total	Ash	Carbon	Sulfur	Oxygen	Nitrogen	Balance
				Pounds			
Refuse	100.0	63.5	25.3	0.5	6.9	0.8	3.0
Air	295.6	0.0	0.0	0.0	68.9	226.7	0.0
Limestone	-	-	-	-	-	-	-
Total in	395.6	63.5	25.3	0.5	75.8	227.5	3.0
Bed	40.3	38.6	1.4	0.06	~0	~0	0.24
Primary cyclone*	11.3	9.8	1.2	0.04	0	0	0.26
Secondary cyclone	15.9	13.7	2.5	0.06	0	0	(0.36)
Bag filters	3.5	3.1	0.3	0.01	0	0	0.09
Flue gas and loss	324.6	(1.7)	19.9	0.33	75.8	227.5	2.77
Total out	395.6	63.5	25.3	0.5	75.8	227.5	3.0

*Primary cyclone recycled through most of run.
Note: Basis–100 lb refuse in.

Source: CONF-751213-3

Sulfur retention (as indicated in Table 11.7) with the Mid-Valley refuse is principally in the fine elutriated ash. The secondary cyclone ash, for example, is
22% of the total solid residue, yet contains 35% of the retained sulfur. The
spent bed conversely is 56% of the solid residue and contains only the same
amount of sulfur. This same effect can be seen in the silt combustion (Table
11.6); however, with the addition of limestone in fine size, the explanation
seems to be the reactivity effect of the finely divided lime particles.

The unburned carbon is likewise found principally in the elutriated fines. This is
especially true for the silt run in which the primary cyclone material, which was reinjected for further burnup, contained 45% carbon. The secondary cyclone ash was
31% carbon while the finest ash retained by the bag filter contained 17% carbon.

The lower initial carbon formed in the Mid-Valley refuse produced correspondingly
lower carbon carryover of 15.7% in the secondary cyclone and 8.6% in the bag filter
residue. The bed contained higher carbon in the combustion of the coarse refuse.
The large slate particles contained visible carbon which could be exposed upon frac-
turing which, perhaps, accounts for some of the excess carbon held in the bed.

Assessment

These waste materials can be burned easily in a fluidized bed combustor to recover
the contained fuel value. Fluidized bed combustors with waste heat boilers located
near the supply of anthracite waste could recover the heating value contained in this
reject material and reduce the demand for expensive and scarce fuels. The burning
of the waste would also somewhat reduce the environmental impact of the refuse
piles, although the ash would remain a disposal problem of somewhat reduced magnitude.

Emission studies from the burning of the wastes indicate that some sulfur capture with limestone or other sulfur acceptor may be necessary to meet emission
standards even though the sulfur content is low in the waste. The specific ques-

tion of how much (or how little) sulfur absorbent will be required with various grades of refuse is still being studied. Anthracite refuse, along with many other types of low-grade fuel, is an acceptable feedstock for fluidized-bed combustors. Such low-grade fuels match the operating advantages of the combustor and represent potential savings of other premium fuels for alternate uses.

CHAR COMBUSTION TO DRIVE A POWER PLANT

Basic Concept

The Consol process for making synthetic crude oil from coal generates char as a by-product which is an undesirable fuel. The char can be burned in a fluidized bed; however, the maximum bed temperature may be limited by the ash agglomeration temperature of the char or by loss of sulfur retention. Instead of using steam as the working fluid, helium was picked, since the maximum working temperature is not limited by tube pressure but can approach the temperature of the bed.

Helium's excellent heat transfer properties assure larger heat transfer rates. Since the maximum working temperature for the helium cycle is higher than that for a conventional steam cycle, the possible cycle efficiency is higher. Since the closed helium loop should be more efficient than raising steam, as much heat as possible should be introduced into the closed cycle by a system of regenerative (between turbine stages) heaters and helium preheaters.

Fluidized Bed Combustion of Char

The Consol process for making synthetic crude oil from coal generates as a by-product approximately 0.26 ton of char per ton of coal fed to the process (5). This amount is in excess of utility fuel requirements of the plant. The char generated by the Consol process has a heating value of approximately 10,000 Btu per pound and represents about one-fourth of the energy value of the coal feed, but its high ash content and low percentage of volatile matter make it an undesirable fuel which is difficult to burn in a conventional furnace. The recent large increase in energy costs enhances the potential value of the char as an energy source and the need to efficiently exploit it.

Char can be burned efficiently without supplemental fuel in a fluidized bed combustor (6). Most of the sulfur content of the char can be removed by adding to the bed dolomite or limestone which will lock up the sulfur as calcium sulfate. However, the maximum temperature of the bed may be influenced by the kinetics of sulfur retention, the formation of NO_x, and the ash agglomeration temperature of the char. (Deactivating sintering of dolomite at high temperatures will not occur before sulfur retention becomes poor for other reasons.)

It has been shown that the ability of dolomite to retain sulfur decreases when the bed temperature increases beyond about 1600°F. Two possible mechanisms are suggested. At higher temperatures the ash may form an inactive layer on the dolomite, closing the pore structure; or a local reducing atmosphere may occur near the bottom of the bed, reversing the SO_2 retention reaction. The latter problem may be solved by increasing the bed height or using more excess air. The amount of retention is a function of the sulfur-calcium kinetics and is not set by equilibrium conditions.

NO_x formation at bed temperatures less than 2400°F is due to oxidation of the nitrogen in the fuel, not direct reaction of nitrogen and oxygen; and the bed is well mixed and local hot spots for direct reaction are not likely to occur. Below 2400°F the other operating conditions of the bed will have a larger effect on NO_x formation than temperature. For example, decreasing particle size will increase NO (the most common NO_x occurring), while increasing operating pressure will lower NO formation. Decreasing excess air will increase the amount of CO formed which reduces the NO.

A temperature dependence for NO_x emissions has been found by several organizations doing research in this field, notably Exxon and the Argonne National Laboratory. However, the increase in NO_x formation with temperature is nearly zero above 1700°F. Depending on the operating conditions and nitrogen content of the fuel, an NO_x level of 250 to 1200 ppm would be expected.

Conventional fluidized beds must be kept below the ash agglomerating temperature of the char (or coal) to prevent collapse of the bed. This phenomenon will occur at about 1900° to 2200°F. Changes in the mechanical design of fluidized beds to remove agglomerated ash clinkers are being applied in developing coal gasification processes. These include the Institute of Gas Technology's U-GAS system, the City University of New York's Fast Fluidized Bed, Bituminous Coal Research's three stage design, and Westinghouse's low Btu gasifier. However, these designs are not necessarily extendable to the high ash content char.

The major limitations on bed temperature would seem to be the loss of sulfur retention by the dolomite and possible excessive ash agglomeration (causing bed collapse). Pilot studies with char can determine the actual maximum desirable bed temperature. In this study a conservative maximum temperature of 1800°F was used.

Helium as Working Fluid

The working fluid selected is helium. A closed helium cycle allows a higher maximum fluid temperature than the more conventional steam system, since the maximum steam temperature is constrained by the pressure requirements of the boiler tubes. No study was made of an open cycle system using combustion products as the primary working fluid. However, since the fluidized bed temperature has a definite maximum, large quantities of excess air would be needed.

With helium as the working fluid, the maximum working temperature can be set by either the allowable minimum temperature difference between the fluid and the combustor (maximum allowable heat transfer area) or by the maximum pressure allowed by materials for the cycle. Since a higher maximum working temperature than with steam is possible, a higher efficiency for the cycle than for a steam plant is possible.

Although several working fluids are possible, helium is the best choice based on the characteristics required for the power cycle. It is chemically inert and has a high specific heat capacity, high thermal conductivity and average viscosity compared with other gases. Only its low density is a drawback, increasing the size and cost of the turbomachinery. However, other studies have shown that the savings in heat exchange equipment costs when using helium far offset the added turbomachinery costs (7)(8)(9).

From this study, a concept was developed for a power plant utilizing fluidized bed combustion of Consol char to drive a closed-cycle helium turbine.

HCl NEUTRALIZATION BY SODIUM CARBONATE FLUIDIZED BED

Principle of Operation

Dow Chemical Company at its Rocky Flats nuclear power plant in Golden, Colorado, has tested the feasibility of a fluid bed incineration process utilizing in situ neutralization of the HCl generated by decomposing PVC in a 2.5-inch diameter quartz laboratory reactor tube. The combustible material was pyrolyzed or combusted in a fluidized bed of sodium carbonate. The HCl generated from decomposition of the chloride-containing plastics reacts with the carbonate to form sodium chloride which remains in the bed and is removed with the ash. This minimizes the concentration of acid gases in the off-gas system and minimizes corrosion of equipment, all at a lower operating temperature (about 500°C).

The data generated on this equipment were used to design a 10 lb/hr pilot plant incinerator.

The test results indicated: Open flame afterburners require about 1000° to 1200°C for complete combustion of the trace hydrocarbons in incinerator flue gas. Oxidation catalysts are satisfactory for use in the afterburner application and should reduce the afterburner temperature to between 500° and 700°C.

A chromic oxide on alumina catalyst is satisfactory at a temperature of about 400°C. Molecular sieve-type materials are satisfactory catalysts for this application but require an afterburner temperature of 500° to 600°C.

The in situ neutralization of the HCl generated from the decomposition of PVC was shown to be more than 90% effective when the waste is reacted in a bed of Na_2CO_3 and the flue gas is passed through a secondary bed of Na_2CO_3. The neutralization proceeds by the following reaction:

$$2HCl + Na_2CO_3 \longrightarrow 2NaCl + CO_2 + H_2O$$

Further neutralization can be accomplished by admixing Na_2CO_3 to the waste prior to extrusion. The major advantage of removing the HCl from the flue gas is that it should minimize corrosion of the incinerator and off-gas scrubbing equipment and should minimize the complexity of the scrubbing equipment. The extruded waste appears to be a suitable form of waste feed for the fluidized bed operation.

The use of 10 to 15% oxygen in the argon or nitrogen used for fluidization offers the best combination of bed-temperature control and low concentration of unburned carbon in the ash. The use of greater oxygen concentrations tends to allow above-bed burning which results in high bed temperatures and poor fluidization. Lower oxygen concentrations result in increased carbon concentrations in the ash.

A reasonable mixture of the waste material is required for the pelletizing operation. Shredding of waste materials is required to obtain this suitable mixture of plastic with the other waste material for the pelletizing operation. The plastic material becomes partially molten during the extrusion operation and it is used to hold the other materials together in a pellet. The pelletized waste material is an acceptable feed for operation of the fluidized bed.

Several shredders were tested and performed successfully with the PVC waste materials. A small low-speed cutter-type of shredder offers the best combination for this application with regard to product size, machine size, cost, etc.

Open-flame incineration is not satisfactory for combustion of leaded gloves because of the formation of lead shot in the ash and the formation of glass by reaction of the leaded ash with refractory materials. The fluid bed incinerator appears to solve these problems by the use of lower incineration temperature (500°C) and because of the ideal temperature control associated with fluid bed operation.

RADIOACTIVE WASTE INCINERATION

Pilot Plant Studies

To overcome some of the incineration problems at Rocky Flats, a program was undertaken by Rockwell International to develop a fluidized bed incineration process to burn combustible scrap containing sufficient plutonium for recovery. A pilot plant was constructed and operated for about two years to develop design information and to obtain operating experience. The pilot plant initially was operated with simulated combustible scrap, and then burned low-level radioactive waste.

This process utilizes in situ neutralization of the acid gases to minimize equipment corrosion and to eliminate the need for an aqueous scrubbing system. An operating temperature of about 550°C eliminates the need for refractory-lined equipment and should produce a more soluble form of plutonium oxide in the ash.

In this process, the waste, which is composed of polyvinyl chloride (PVC), polyethylene, and paper, is chopped in a low speed, cutter-type shredder. The waste is then fed to the incinerator by a constant pitch, tapered screw conveyor at a rate of about 20 lb/hr. During operation, the shredded waste is introduced into the primary combustion chamber under the surface of a fluidized bed of granular sodium carbonate (Na_2CO_3).

The primary combustion chamber operates at a temperature of approximately 550°C. Hydrogen chloride generated by PVC decomposition reacts with the bed material to form sodium chloride (NaCl). The fluidizing gas is a mixture of nitrogen (N_2) and air. The amount of air used is limited so that pyrolysis of the combustibles occurs and the operating temperature of 550°C is maintained.

Particulates generated in this vessel are then partially removed by a cyclone, and the products of pyrolysis are introduced into the bottom of a catalytic fluidized bed with enough air to ensure complete reaction. Flue gas dust generated in the afterburner at 550°C is fed to a cyclone and a sintered metal filter for removal of the entrained solids. The gas stream then enters an air jet ejector. This ejector provides the motive force for gas flow through the reactors and filtering equipment and also cools the gas stream by dilution. This stream is blended with room air and passes through one stage of high efficiency particulate air (HEPA) filtration before it leaves the incineration room. It then passes through four stages of HEPA filtration before reaching the outside atmosphere.

Based on data generated by this equipment, a demonstration unit with a capacity of 82 kg/hr (180 lb/hr) is being constructed for combustion of transuranic (TRU) waste. A description follows.

Demonstration Plant Design

The fluidized bed incineration process used for combustion of transuranic (TRU) waste differs from that of the pilot plant in the method of heat removal and the method of producing the motive force for gas flow through the equipment. In the pilot unit, the heat of combustion was removed by blending a large amount of room air with the flue gas.

In the demonstration plant there are mechanical blowers which will be used rather than the air jet ejector. This will reduce both equipment cost and operating cost. The design for a heat release rate of 1,500,000 kJ per hour (1,500,000 Btu per hour), corresponds to a capacity of about 82 kg per hour (180 lb per hour). A flow diagram of the process is presented in Figure 11.5.

Waste will be received in 200 liter (55 gallon) drums. The waste will be introduced from an air lock box (in an air lock room) into the feed preparation area. Large pieces of metal in the waste will be removed by hand sorting to prevent damage to the first shredder. This metal will be collected and drummed for disposal. Combustible waste from the hand sorting will then be fed to a low-speed, cutter-type shredder for coarse shredding.

Associated with the combustibles will be small pieces of metal that were undetected during hand sorting. This tramp metal will be shredded along with the waste.

It is intended that discharged waste from the first shredder be conveyed to an air classifier. This classifier will allow collection of the heavy tramp metal in a glove-ported receiver. Periodically the metal can be bagged out for disposal.

Light combustible waste shall then be conveyed in an air stream to a shredder that has narrower and closer spaced blades. The waste will be shredded again and be of a suitable particle size for feeding to the pyrolyzing fluidized bed.

To introduce the waste to the first fluidized bed, a constant pitch screw will be used. This should compact the waste and feed it into the bottom of the fluidized bed. The screw will be variable speed and controlled by the temperature in the catalytic fluidized bed. As the temperature in the catalytic fluidized bed increases, the feed rate of waste to the first fluidized bed will be reduced. As the temperature decreases, the feed rate is increased.

FIGURE 11.5: FLOW DIAGRAM FOR DEMONSTRATION PLANT FLUIDIZED BED INCINERATOR

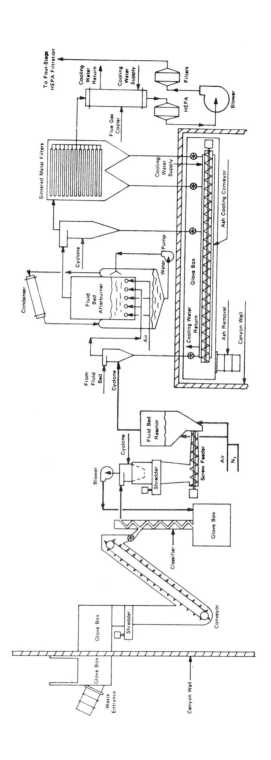

Source: RFP-2471

The first fluidized bed will accept the waste and a mixture of air and nitrogen. The temperature of the bed is to be controlled by the amount of air fed to the unit. If more air is required to maintain bed temperature, the amount of nitrogen will be reduced proportionately to maintain a constant flow of fluidizing gas to the unit. As with any fluidized bed, particulate matter will be entrained in the gas stream leaving the unit.

To remove a large portion of particulates, a cyclone is to be installed between the two fluidized beds. This elutriation of solids is the mechanism for removing chlorides neutralized by the sodium carbonate (Na_2CO_3). The cyclone will then pass the combustible gases and a small portion of the solids to the catalytic fluidized bed.

The second fluidized bed shall contain a granular catalyst of chromium oxide on an alumina base. This catalyst will promote oxidation of the pyrolyzed gases at about 550°C. Sufficient air is then added to the bed to provide an excess of oxygen and to ensure uniform and complete fluidization. The combustion products and entrained catalyst are discharged from the afterburner to the dust collection equipment.

The gas stream from the catalytic fluidized bed goes through two processing steps to remove the entrained solids. The first step involves a cyclone where 75 to 85% of the solids will be removed.

The second unit is to be a bank of sintered metal filter tubes that will remove remaining dust down to a particle size of one micron. Each set of tubes in this bank shall be sequentially back-blown with air or nitrogen to remove the build-up. Gas stream temperature will be maintained above 400°C by insulation of the dust collection equipment and process lines. As a result, should a process upset occur, the hydrocarbons produced by pyrolysis will not condense and plug the filter tubes. The gas stream could then be cooled for further processing.

The discharge temperature from the filters is calculated to be approximately 500°C. Using a water-cooled heat exchanger, this gas stream will be cooled to about 50°C to prevent damage to the equipment downstream.

The motive force for this system shall be supplied by centrifugal blowers. The blowers will be sized to induce enough draft at flowing conditions to return the process pressure back to atmospheric pressure. Room air is to be fed to the inlet of these blowers to control the draft in the process. The pressure sensor for this control loop will be located in the first fluidized bed, and the controller can be set at about 19 cm water column (7 inch) to 25 cm water column (10 inch) below atmospheric pressure.

Blower discharge shall be filtered through a set of HEPA filters before leaving the canyon. The process gases and the canyon air then will be ducted to a four-stage HEPA plenum before being discharged to the atmosphere. The fluidized bed incineration system will occupy an area of approximately 3,000 square feet and will require monitoring of at least 75 data sensing points in process equipment. A computer will monitor the process and store variables for regression analysis.

FLUIDIZED BED SOLIDIFICATION OF NUCLEAR WASTES

Emergency Fluidization and Bed Removal

Work is now in progress at the Idaho Chemical Processing Plant (ICPP) of Allied Chemical Corporation to obtain process and equipment design data for fluidized bed solidification of the high-level liquid wastes generated during the reprocessing of commercial light water reactor (LWR) fuels.

The basic conceptual process design for a commercial fluidized bed solidification facility is similar to an existing Waste Calcining Facility (WCF) at the ICPP; process and equipment improvements being designed into the New Waste Calcining Facility (NWCF) are also directly applicable to a commercial facility.

Chemical dissolution of fuel (cladding and heat) at the ICPP and the chop-leach process used and planned for near-term commercial reprocessing plants result in significant differences in the volumetric heat generation rates of the calcined solids. The calcined solids generated at the ICPP have heat generation rates in the range of 5 to 50 Btu/hr-ft^3; calcined solids from the fluidized bed solidification of LWR first cycle raffinate will have heat generation rates in the range 4,400 Btu/hr-ft^3 (5 years out of reactor) to 22,000 Btu/hr-ft^3 (1 year out of reactor).

One concern during fluidized bed solidification of LWR reprocessing wastes is the ability to remove decay heat if the fluidized bed should collapse because of accidental loss of fluidizing air. A collapsed fluidized bed in the calciner that is in operation at the ICPP has not and could not result in any significant heat-up because of the low volumetric heat generation rate. However, collapse of the fluidized bed in a commercial calciner, assuming no remedial action was taken, could result in significantly higher temperatures.

The rate of bed temperature rise in a collapsed commercial calciner bed is sufficiently slow to permit a deliberate response. The heatup rate depends primarily on the decay-heat generation rate and to a lesser extent on the bed diameter, height, and thermal conductivity. With no heat dissipation it would take 0.5 and 3 hr, respectively, with 1- and 5-year-cooled wastes, for the center of the bed to reach 1300°F, the temperature at which cesium begins to volatilize. Thus, a "hair-trigger" response to a loss-of-fluidization is unnecessary.

Cooling of the calcined solids by redundant fluidizing air systems and removal of the solids from the calciner are considerations as primary safeguards in the event of a collapsed bed. Fluidizing air at normal velocities (1 to 4 ft/sec) will maintain the solids at 900°F (the nominal calcination temperature).

Fluidizing air provided by the following set of systems would essentially eliminate the possibility of loss of fluidizing air:

[1] regular fluidizing air blowers with in-line spares,
[2] the plant compressed air system,
[3] an emergency compressed or liquid air supply, and
[4] emergency fluidizing air blowers powered by gasoline or
     diesel engine.

The design of the fluidizing air system for a specific plant would depend on the overall plant design and safety analysis, and on specific system reliability requirements.

Redundant systems should be provided for removing the bed if removal is required for maintenance or long-term cooling. Bed removal should be at the discretion of the operators. Candidate bed removal systems are as follows:

[1] normal bed (product) removal system consisting of two or three removal lines each with capability of removing plugs,

[2] bed discharge (dump) lines through the fluidizing air distributor plate, and

[3] emergency bed removal by a "vacuum" system extending into the bed above.

The consequence of a collapsed bed, even if no actions were taken, would not result in a hazard potential to the public that would be greater than that in routine plant operation. Process and cell off-gas cleanup systems would still prevent the release of volatile or particulate matter, above acceptable limits, to the atmosphere.

A sintered (or partially molten) bed would be an operating problem rather than a safety or environmental hazard. The wall temperature of a partially insulated or uninsulated calciner vessel would be well below the melting temperature of the wall, and decay heat, even for 1 year old waste, could be dissipated from the vessel wall by natural convection and thermal radiation to the cell off-gas and the cell wall.

Reliable redundant fluidizing air systems for cooling can be designed by application of existing plant engineering and quality assurance procedures. Concepts for bed removal are being evaluated, both theoretically and experimentally. They include a product system, a bed discharge system, and a backup bed discharge system.

Bed removal can and should be at the discretion of the operators. The automatic discharge of the bed is not needed for safety (assuming reliable emergency fluidization); the premature discharge of the bed is undesirable from an operational viewpoint.

Other emergency cooling or bed removal concepts that could be considered if increased redundancy is desired are a butterfly distributor plate, diluent addition of aluminum powder, helium addition, weeping distributor plate, water quench, and elutriation.

The emergency compressed air fluidizing concept and the vacuum (backup bed-discharge) concept are being evaluated on a reduced scale in a pilot-plant calciner to demonstrate the feasibility and operability of the two concepts. In addition, the solids discharge rate realized using the vacuum bed-discharge system will be measured and the effects of operating parameters evaluated.

COMBUSTION DISPOSAL OF MANURE WASTES

Generation of Agricultural Wastes

Though agricultural wastes are much less conspicuous to the general public than urban and industrial refuse, the amount generated far exceeds that of any other category. Estimates indicate the total production of agricultural wastes is about 2 billion tons annually, accounting for almost 60% of the total solid wastes generated in this country.

Of this 2 billion tons produced, 1.1 billion tons is animal manure, 0.4 billion tons is animal liquid waste, and 0.5 billion tons comprise miscellaneous meat packing, forestry, and farm wastes.

Approximately half of the 1.5 billion tons of animal waste is produced in large cattle and hog feedlots, large dairy operations, and chicken and turkey operations where 100,000 or more fowl are grown annually. Such enterprises are often adjacent to urban-industrial-suburban complexes, which further complicates the pollution problem and necessitates the development of better and faster disposal methods.

Described here is the combustion of animal manures in a fluid bed reactor to determine if the combustion products could be utilized, with a brief comparison with the results for combustion in a rotary kiln.

Materials Studied: The materials used for this study were samples of manures obtained from different animal–raising operations. Samples of horse, hog, dairy cow, broiler chicken, egg-farm chicken and feedlot cattle manures were obtained. Analyses of the oven-dried manure, including the percent moisture of the as-received manures, are given in Table 11.8 which shows how the composition of the different manures varies with the animal source and the character of feed.

TABLE 11.8: ANALYSES OF OVEN-DRIED MANURES*

Type of Manure	C	S	N_2	K_2O	P_2O_5	CaO	Al_2O_3	SiO_2	Moisture (as received wet manure) (%)
Cattle (feedlot)									
#1	39.0	0.6	0.3	2.4	4.1	4.6	0.5	5.1	70
#2	43.2	0.5	0.3	0.8	1.7	1.4	0.9	5.8	75
#3	16.6	0.4	0.1	1.3	0.9	3.6	6.5	40.7	25
Dairy cow	37.4	trace	0.2	1.7	1.4	1.0	0.9	15.0	80
Chicken									
Broiler**	34.0	ND	3.4	2.3	2.5	1.7	0.4	3.2	21
Egg farm***	30.4	0.3	3.1	2.5	4.7	10.5	0.8	2.4	74
Hog	35.0	trace	0.3	1.5	4.2	3.6	1.0	14.9	67
Horse	ND	ND	0.1	1.8	1.2	1.4	1.1	31.0	50

Note: ND = not determined

*Oven dried at 90°C for 48 hours.
**Contained wood chips, used as litter.
***Contained no wood chips.

Source: PB 206 892

Heats of Combustion of Manures

Preliminary batch fluid bed tests using air as the fluidizing gas had indicated that the dry manure would burn and possibly support combustion. Thus, if the heating value of the manures was sufficient to support combustion, then the reaction in the continuous fluid bed might be self-sustaining, eliminating the need of additional heat from the reactor furnace.

Therefore, heat of combustion determinations were made for each oven-dried manure in an adiabatic oxygen bomb calorimeter. Results of the determination presented in Table 11.9, show that the values range from 2,520 to 7,810 Btu/lb (1,400 to 4,340 calories per gram). Based on these data, all but the #3 cattle manure and possibly the egg-farm chicken manure would support combustion and burn at a high enough temperature to be self-sustaining.

TABLE 11.9: HEAT OF COMBUSTION FOR OVEN-DRIED* MANURES

Type of Manure	Btu/lb	Cal/g
Cattle (feedlot)		
#1	6,410	3,560
#2	7,810	4,340
#3	2,520	1,400
Dairy cow	6,500	3,610
Chicken		
Broiler**	6,800	3,780
Egg farm***	4,370	2,430
Hog	6,100	3,390

*Oven-dried at 90°C
**Contained wood chips
***Contained no wood chips

Source: PB 206 892

Continuous Fluidized Bed Tests

The continuous fluid bed tests were made in two reactor systems consisting of fluid bed reactors, a screw-type feeder, a gaseous cyclone for dust removal, a condenser for water removal, an air preheater, and other apparatus to collect the products of reaction.

The first system, had a 1.75-inch diameter reactor with the screw feeder on top. The feed was transported by the screw to a vertical tube which fed to the inside bottom of the reactor. This arrangement worked fairly well, but periodically the tube would plug, halting the reaction. To remedy this problem, and study the operation of a larger reactor, the second system was constructed using a 2.5-inch diameter reactor with the screw feeder feeding directly into the side of the reactor.

In all tests the manure feed was dried at 90°C and crushed to pass 10 mesh. The reactor was heated to 400°C in a resistance furnace and fluidizing air, preheated to 250°C was used prior to starting the feed and throughout the tests. The feed rate and airflow rate were calculated prior to the test so that complete combustion could be maintained. After ignition, the power to the reactor was turned off and the temperature rose to the operating temperature of 770° to 860°C, staying there until completion of the test.

The fine residue from combustion was removed from the reactor with the exit gas and separated from the gas in the cyclone. The coarse residue remained in the smallest reactor but was discharged to a collection flask from the bottom of the larger reactor. The exit gas, after separation from the residues, passed through the condenser to remove the water resulting from the combustion. In some cases, the exit gas was also bubbled through H_2SO_4 solution to remove the ammonia produced in the reaction before venting to the atmosphere.

A portion of the exit gas was monitored continuously for oxygen content using an oxygen analyzer. By monitoring the exit gas and adjusting the inlet air when needed, the oxygen content of the exit gas was maintained at 1 to 3%, insuring an oxidizing atmosphere in the reactor. The exit gas was also periodically analyzed for CO_2, CO, and O_2 using an Orsat gas analyzer.

Results of the fluid bed combustion tests are compiled in Table 11.10. The fine and coarse residues were combined for analysis. From 75 to 80% of the combined residues was collected as fines in the cyclone. In all tests, no CO was detected in the exit gases.

TABLE 11.10: COMPILATION OF CONTINUOUS FLUID BED TESTS ON OVEN-DRIED* ANIMAL MANURES

Type of Manure	Reactor Used ID (inch)	Feed Rate (g/hr)	Airflow Rate** (ft³/hr)	Maximum Reactor Temp. (°C)	Residues								Exit Gas	
					Wt (g/hr)	C	S	K₂O	P₂O₅	CaO	Al₂O₃	SiO₂	CO₂ (%)	O₂ (%)
Cattle														
#1	1.75	320	60.0	850	80	8.7	1.7	9.0	16.1	24.0	1.3	14.7	16.8	1.8
#2	2.5	970	146.0	860	75	8.5	1.4	13.5	14.9	10.4	2.4	30.8	15.3	2.1
Chicken														
Broiler	1.75	270	54.0	770	45	7.1	1.2	8.0	14.0	5.5	3.2	31.3	ND	ND
Egg farm	2.5	800	90.0	780	180	8.9	1.1	8.4	15.9	42.3	0.9	2.6	13.5	3.0
Hog	2.5	600	75.0	850	175	8.2	2.6	12.9	14.2	15.1	2.1	35.2	14.5	1.5

Note: ND = not determined.

*Oven-dried at 90°C.

**Dry air at STP–air heated at 250°C before entering reactor.

Source: PB 206 892

Rotary Kiln Comparison: Comparison combustion tests were made in a small gas fired rotary kiln. An important difference between the rotary kiln and the fluid bed tests was that wet manure was fed to the rotary kiln, whereas dry manure feed was used in all the fluid bed tests.

In the rotary kiln tests, the kiln was heated to approximately 600°C by means of a natural gas burner prior to feeding the wet as-received manure. After the feed was started and steady burning of the manure was observed, the burner was throttled down to a minimal flame and maintained as such until completion of the test.

As the wet manure entered the kiln it dried and then burned as it traveled through the kiln. The residue was discharged at the burner end of the kiln and collected. A very minor amount of the residue was removed with the exit gas. The exit gas was periodically analyzed for CO_2, CO, and O_2 using an Orsat gas analyzer. In all tests, no CO was detected in the exit gases.

Assessment

The fluid bed combustion tests showed that the oven dried animal manures will support combustion and are self-sustaining when using air preheated to 250°C. In the combustion reaction, over 75% of the weight is lost, and this reduction increases to 90% or better when basing the weight loss on the wet as-received manures. These tests also showed that a reduction in volume of as much as 85% is obtained when the manure is burned.

The nitrogen in the residues is less than 0.01%, and essentially all the nitrogen in the feed is removed as ammonia in the effluent gases from the bed. Analysis of the H_2SO_4 absorption solutions used in some of the tests showed that greater than 95% of the original nitrogen can be recovered.

The potassium and phosphorus in the feed to both reactors concentrates in the residues. Tables 11.8 and 11.10 show that the residues contain from 8 to 14% K_2O and from 13 to 16% P_2O_5 with the exception of the high silica Greeley manure.

Leaching tests of residues showed that 80 to 85% of the potassium was soluble in 2 hours while only 40% of the phosphorus was soluble in a week-long test. Although the water solubility of the phosphorus was low, results using the standard citrate-insoluble phosphoric acid analysis showed that over 90% of the phosphorus present in the residue was available.

The results of these tests indicated that the residues could be used as a nitrogen-free-type fertilizer containing significant amounts of potassium and phosphorus. The remaining portion of the residues was mostly lime, silica and alumina, which have possible uses as soil conditioners.

Dry manure was used as feed in the fluid bed tests primarily because of the size and construction of the feeding system. In large fluid bed systems, methods of feeding the wet manure probably can be devised, but in order for the system to function properly and the combustion to be self-sustaining, sufficient heat must be available to dry the feed prior to burning.

To estimate whether the heat generated by combustion is sufficient to dry the wet manure and also preheat the air to 250°C, a heat balance calculation was made based on the data from burning cattle manure. Results are shown in Table 11.11.

The data in Table 11.11 show the heat balance for feeding the manure into the fluid bed reactor in two ways—[1] feeding wet manure directly into the reactor and [2] predrying the wet manure with the exhausted combustion gases and then feeding the dry manure into the reactor. The calculated results indicate that sufficient heat is not available to dry and burn the wet manure if it is fed directly into the fluid bed.

TABLE 11.11: ESTIMATED HEAT BALANCE FOR FLUID BED COMBUSTION OF FEEDLOT CATTLE MANURE

Phase of the Operation	Wet Manure to Reactor Required	Available	Predried Manure to Reactor Required	Available
Heat into system				
Heat of combustion	–	19,000	–	19,000
Heat in preheated air	–	2,000	–	2,000
Heat in end products				
Residue	296	–	296	–
Unburned carbon in residue	934	–	934	–
Exit gases	6,960	–	6,960	5,220
Heat to preheat air	2,670	–	2,670	–
Heat for water in wet manure				
To evaporate water	7,250	–	9,670	–
To bring water vapor to 850°C	5,150	–	–	–
Heat to bring manure solids to 850°C	742	–	742	–
Total heat	24,002	21,000	21,272	26,220
Net Heat	3,000	–	–	4,950

Notes: Calculations based on 1 minute of operation.
    Feed rate: 5.33 g/min oven-dried #1 cattle manure.
    Reference temperature: 25°C.
    Reaction temperature: 850°C.
    Fluidizing gas flow: 1.0 ft^3 air STP.
    Other data taken from Table 11.10.

Source: PB 206 892

However, the available heat can be more efficiently used if processing of the manure is carried out so that the wet manure is dried before fluid bed combustion. Under these conditions the waste heat in the exit gases can be used to dry the wet manure and the heat would not be required to raise the water vapor to the reaction temperature of 850°C.

When this system is used the data indicate that sufficient heat, with an excess of about 5,000 cal/min, is available from combustion of the manure to provide for drying the wet manure and preheating the air for combustion. For the combustion to be maintained satisfactorily, manures having lower heating values than those used in the calculation might require supplemental heat to predry the manure and preheat the air.

REFERENCES

(1)   MacCartney, J.C., and R.H. Whaite, *Pennsylvania Anthracite Refuse, A Survey of Solid Waste from Mining and Preparation,* U.S. BuMines Information Circular No. 8409, 1969, 77 pp.

(2) Cooke, M.J. and N. Hodgkinson, *The Fluidized Combustion of Low-Grade Materials,* Proceedings of Fluidized Combustion Conf., London, September 1975, Inst. of Fuel Symposium Series No. 1 (1975), pp. C2-1–C2-9.

(3) Norman, P., "Combustion of Colliery Shale in a Fluidized Bed," *Energy World,* July 1975, No. 18, pp. 2-4.

(4) Coates, N.H. and R.L. Rice, *Sulfur Dioxide Reduction by Combustion of Coals in Fluidized Beds of Limestone,* AIChE Symposium Series, vol. 70, No. 141, 1974, pp. 124-129.

(5) The Ralph M. Parsons Company, *1969 Feasibility Report Consol Synthetic Crude Production,* (Los Angeles, Calif.: The Ralph M. Parsons Company, 1969).

(6) McLaren, J. and D.S. Williams, "Combustion Efficiency, Sulfur Retention, and Heat Transfer in Pilot Plant Fluidized Bed Combustors," *Journal of the Institute of Fuel,* (August 1969): 303-308.

(7) Robson, F.L. et al., *Technological and Economic Feasibility of Advanced Power Cycles and Methods of Producing Nonpolluting Fuels for Utility Power Stations* (East Hartford, Conn.: United Aircraft Research Laboratories, 1970), p. 224ff.

(8) Biancardi, F.R., et al., *Progress Report on Studies of Advanced Nuclear Power Generating Systems,* (East Hartford, Conn.: United Aircraft Research Laboratories, 1963).

(9) Keller, C., "The Use of Closed-Cycle Helium Gas Turbine in Atomic Power Plants," *Escher Wyss News,* 39(1) (1966).

PROPRIETARY DISPOSAL SYSTEMS
AND SPECIAL DESIGNS

INDUSTRIAL WASTE DISPOSAL SYSTEMS

Incineration of Liquid Sludge Containing a High Proportion of Salts

U.S. Patent 3,921,543 describes a process for burning water-containing wastes, particularly clarifier sludges which have a high salt content, particularly sodium chloride, in a fluidized-bed furnace operated with an oxygen-containing gas phase wherein the turbulent gas has a mean velocity of parallel flow of at least 3 meters/second at the surface of the fluidized bed; the waste is incinerated in a bed of inert solids with the bed heated to a temperature above the melting point of the salts. The superficial velocity and/or the bed temperature is adjusted to ensure that the predominant part of the salt is discharged with the exhaust gases.

In this connection the mean velocity of parallel flow of the fluidizing gas (also referred to as the superficial velocity) is the mean velocity of parallel flow in the empty fluidized bed furnace without its solids bed. The turbulencing gas is the gas which is produced by the incineration of the waste with the addition of fuel and oxygen containing fluidizing gas.

The materials fed to form the bed consist of materials which either do not react or else form compounds with the fed waste, with components thereof or with the combustion products which have melting and sintering points above the operating temperatures of the fluidized bed furnace.

Particularly suitable bed materials are iron oxide granules having an average particle diameter of 0.2 mm to 4 mm and a bulk density above 2.5 kilograms per liter, or corundum having about the same particle size. These oxides are generally used only when a new plant is put into operation because during the incineration of sludge the inorganic substances fed in with the sludge are transformed into particularly oxidic compounds or mixed oxides of iron, calcium and manganese and these compounds constitute a so-called chemical bed material which may be used alone or in a mixture with the starting material.

Such a bed material has excellent properties and because it is produced as a by-product in many industrial plants it is also eminently suitable as a starting material for new plants or for a new operation cycle.

In the process, the mean velocity of parallel flow of the turbulencing or fluidizing gas and the temperature of the fluidized bed should be selected in dependence on the resulting common salt vapor pressure. Thus common salt is made to be discharged with the exhaust gases approximately at the rate at which it enters the fluidized bed furnace with the water-containing waste. The operating conditions can be determined most simply by observing the fluidized bed through a sight glass. If these is an agglomeration of bed material in certain parts of the bed surface, the speed of the turbulencing gas and/or the bed temperature is to be increased.

Satisfactory operation could also be determined by an analysis of the bed material if a chemical bed material is used. Any increase of the common salt content will then suggest that the feed rate of the fluidizing gas and/or the bed temperature is too low.

A third method of determination is recommended if the exhaust gases are scrubbed to remove common salt. The salt content of the spent scrubbing water and the common salt content of the sludge are analytically determined and compared.

To prevent an undesired enriching of salts as a result of inevitable small fluctuations of the flow rate of the turbulencing gas and of the temperature of the fluidized bed, the temperature of the fluidized bed is suitably adjusted to be about 30° to 50°C above the desired value which has been determined.

The temperatures of the fluidizing bed are in the range of 900° to 1100°C. In case of high common salt contents it may be desirable to increase the mean velocity of parallel flow of the turbulencing gas to values above 4 meters/second.

Incineration of Wastes Containing Alkali Metal Chlorides

For incineration of combustible waste streams or sludges which have a high chloride content, U.S. Patent 3,907,674 describes the introduction into a fluid bed reactor, operating on these wastes at an elevated temperature and in the presence of sulfur-containing compounds, of a source of silicon dioxide and a metal oxide. They react with the alkali metal-containing compounds present in the reactor to prevent formation of sticky alkali metal silicate glasses or, if such glasses are formed, to convert them to high-melting crystalline silicate compounds.

More specifically, SiO_2 of average particle size –325 mesh and at least one metal oxide from the group consisting of CaO, MgO, Al_2O_3 and Fe_2O_3 may be added directly to a chloride-containing waste stream prior to introduction into the fluidized bed; or alternatively, the silica and metal oxide may be separately injected into the bed. As an example, pulverized sand and slaked lime, $Ca(OH)_2$, may be employed. Instead of lime, or in substitution for a part of the lime, one or more of the metal oxides MgO, Al_2O_3 or Fe_2O_3 may be used as devitrifiers either to prevent formation of sticky silicates or to react with such silicates to form high-melting nonsticky silicates. Various sources of these oxides may be used as long as they are in a suitable state of subdivision. In some cases it is possible to choose the method of concentrating the waste so as to use substances which will later be useful in the incineration operation. For example, a diatomaceous filter aid could be used for a filtration step which would later serve as a source of at least a part of the silica in the incineration step. Likewise, an alum floc [Al(OH)$_3$] could be used to collect oil droplets and this would serve as a source of at least a part of the Al$_2$O$_3$ needed to protect the bed against defluidization.

A large proportion of the alkali metal chloride that is present in the incoming feed will be volatilized by the high temperature in the reactor and will leave the reactor as components of the exhaust gases. This process is particularly concerned with that part of the alkali metal chlorides which, rather than leaving the reactor promptly in a gaseous phase, tends to react with the alkali metal present in the reactor to form deleterious products. It is found that some of the alkali metal in the incoming feed tends to react with sulfur-containing compounds to form undesirable alkali metal sulfates in the bed.

Concentration of Waste Sludge for Fluidized Bed Incineration

There is a real need for a process for the incineration of industrial waste streams which not only effectively disposes of the potentially polluting materials in the waste stream, but also relies for its energy requirements primarily upon the combustible materials within the waste stream itself.

Accordingly, U.S. Patent 3,926,129 describes a method for utilizing the combustible constituents in an industrial waste stream in an incineration process to satisfy the energy requirements of the incineration process.

Then, in a fluid bed process for the incineration of industrial waste streams, the heat in the exhaust stream from the fluid bed reactor is efficiently utilized while at the same time providing adequate treatment of all exhaust streams to remove pollutants therefrom without imposing on the fluid bed reactor a recirculating load of inert ash.

A pair of gas-to-gas heat exchangers operates on the exhaust gas stream of the fluid bed incinerator. The first of these heat exchangers operates to heat a stream of air (and to cool the reactor exhaust gases) which is then employed as the fluidizing and combustion gas for the fluid bed reactor.

The second of these heat exchangers operates to heat an air stream which is then conducted to a venturi scrubber-evaporator, in which weak feed liquor is fed into the venturi with the hot gas. Evaporation occurs and the underflow from the venturi scrubber-evaporator is a strong liquor having an increased concentration of combustible substances therein which is then fed to the fluid bed reactor.

The exhaust gas from the venturi scrubber-evaporator may be discharged to atmosphere. The exhaust gas stream of the fluid bed reactor after traversing the two heat exchangers is directed to a venturi scrubber which traps and separates particulate material in the gas stream which can then be discharged to atmosphere.

Disposal of Nitrogenous Waste Materials

According to U.S. Patent 3,916,805, a nitrogenous material is disposed of without substantial nitrogen oxide formation by introducing the nitrogenous material into a fluidized bed reduction zone containing a catalyst, preferably nickel, for the reduction of NO_x, decomposing the nitrogenous material under reducing conditions in the presence of catalyst in the reduction zone, and withdrawing gaseous products of partial oxidation and decomposition from the reduction zone.

Oil shale or low sulfur coal can also be treated according to this process. In this regard, finely ground oil shale can be partially burned in a single stage process in order to obtain a combustible gas containing hydrogen and carbon monoxide and having a very low NO_x content. Where maximum production of energy is required, as for example in steam generation for an electric power plant, oil shale can also be burned. The oil shale is crushed to a suitable particle size and may be fed to the reactor by known means for feeding pulverized solids.

Combustion is carried out in two stages. The nitrogenous material to be disposed of, auxiliary fuel, e.g., propane, where required, and a substoichiometric quantity of oxygen (usually in the form of air) are admitted to a first fluid bed stage, where partial combustion takes place in the presence of a catalyst. Then secondary oxygen (also usually in the form of air) is added in an amount in excess of that required to complete the combustion of the gaseous products formed in the first stage. The overall result is complete decomposition of the nitrogenous material with the formation of a stack gas containing only small and environmentally acceptable amounts of nitrogen oxides, carbon monoxide and hydrocarbons.

The temperature in the reduction zone is preferably from about $1600°$ to about $2000°F$. Efficient catalytic reduction of nitrogenous decomposition products to elemental nitrogen takes place only at temperatures above about $1200°F$, and minimum nitrogen oxide formation is achieved when the reduction zone temperature is above about $1600°F$.

On the other hand, the temperature in the oxidation zone should not exceed about $2500°F$, in order to avoid or at least minimize conversion of atmospheric nitrogen in combustion air to nitrogen oxide; when a single fluid catalyst bed encompasses both the oxidation and reduction zones, the maximum temperature in the reduction zone is also about $2500°F$. Actually, it is ordinarily preferred to operate at temperatures below about $2000°F$, since this permits a wider choice of materials for constructing the reactor and associated equipment without any sacrifice of efficiency in nitrogen oxide abatement.

Ordinarily the decomposition of a nitrogenous material is not normally self-sustaining, even though some heat is liberated in the partial oxidation of carbon and hydrogen constituents of an organic nitrogenous material. This is due primarily to the fact that large amounts of water are ordinarily fed to the reactor as a slurry medium for the nitrogenous material, and the heat requirements for evaporating this water exceed the amount of heat liberated by the partial oxidation of carbon and hydrogen in the nitrogenous material. The large amounts of water are particularly desirable in the case of explosives to ensure safe handling.

However, some nitrogenous materials, notably oil shale and nonexplosive industrial nitrogenous waste materials, can be handled in the dry state. No auxiliary fuel is required in those instances in which the heat liberated by the partial oxidation of the nitrogenous material is sufficient to maintain the reaction temperature within the desired range.

Organic Nitrogen Compounds

U.S. Patent 3,888,194 describes a method for incinerating industrial waste containing organic nitrogen compounds, by using a fluidized bed furnace, in which industrial waste, together with materials containing a great amount of carbon, such as pulverized coal, heavy oil or the like, is charged in a furnace to thereby fluidize the waste and the carbon-abundant material together with fluidizing medium and air.

The industrial waste is then burned, while the interior of the fluidized bed (or fluidized layer) is maintained in a reducing atmosphere, whereby the nitrogen oxides (NO_x) produced from the organic nitrogen compounds contained in the waste are reduced during the combustion to thereby obtain nitrogen gas (N_2), thus preventing the production of nitrogen oxides.

The fluidized bed is provided by blowing air or gas into a furnace from its bottom and fluidizing the fluidizing medium, a solid particle layer which is supported on a perforated plate within a furnace. In general, silica-base particles (primary constituent SiO_2) such as river sand, which is chemically inactive, are used as the fluidizing medium, the particle size thereof being 0.1 to 2.0 mm. If sulfur and chlorine are contained in the waste and desired to be directly fixed and absorbed within the fluidized bed, it is preferable that chemically active cement clinker particles be used as the fluidizing medium. The composition of cement clinker particles is, for instance, 60 to 70% CaO, 6 to 8% Fe_2O_3 and 3 to 6% SiO_2.

The temperature of the fluidized bed is normally in the range of 750° to 850°C. In this respect, such a temperature is much lower than the temperature (over 1000°C) at which nitrogen oxides are formed by the reaction of nitrogen with oxygen contained in air, as is the case with industrial furnaces, such as boilers. Furthermore, at such a low temperature, the quantity of nitrogen oxides produced according to the reaction of nitrogen with oxygen in air is negligible.

What is of consequence in this respect is the production of nitrogen oxides due to the combustion of organic nitrogen compounds contained in the waste, because nitrogen oxides may be produced in great quantity even at a temperature as low as 750° to 850°C.

Vigorous reaction may be achieved in the fluidized bed furnace, because the fluidizing solid particle layer consists of particles of a size from 0.1 to 2.0 mm and thus has an extremely large contact surface area at the time when the particle layer stands still (e.g., the height of the layer is 500 to 1,000 mm). Accordingly, even waste having poor combustibility, like a dehydrated mudcake (water content, 70 to 90%) may be burned at a very low excessive air ratio (1.1 to 1.3).

The reactions which take place within the fluidized bed incineration furnace according to the process are summarized as follows. The following reactions take place for the combustible matter and the added carbon components, within the reducing fluidized bed which is maintained at a temperature of not less than $500°C$:

$$2C + O_2 \longrightarrow 2CO$$
$$C + H_2O \longrightarrow CO + H_2$$
$$C_nH_m \longrightarrow aCH_4 + bC_n'H_m'$$

where C_nH_m represents hydrocarbon groups contained in the combustible materials, part of which is subjected to dry distillation and decomposition within the fluidized bed under a reducing atmosphere to thereby produce methane (CH_4) and the like.

Carbon monoxide, hydrogen and methane which are produced from the reactions maintain the fluidized bed in a reducing atmosphere, whereby nitrogen oxides produced simultaneously with the production of the gases are reduced according to the following reactions:

$$2NO + CO \longrightarrow N_2O + CO_2$$
$$N_2O + CO \longrightarrow N_2 + CO_2$$
$$2NO + 2H_2 \longrightarrow N_2 + 2H_2O$$
$$CH_4 + 4NO_2 \longrightarrow 4NO + CO_2 + 2H_2O$$
$$CH_4 + 4NO \longrightarrow 2N_2 + CO_2 + 2H_2O$$

The quantity of CO produced within the fluidized bed according to the above reactions is larger than that of nitrogen oxide. Excessive CO, H_2, $C_n'H_m'$, CH_4 and the like are oxidized by means of the secondary air in a space over the upper surface of the fluidized bed and are converted into CO_2 and H_2O, harmless gases.

When incinerating sewage mud having a water content of 78%, solid matter of 22% and nitrogen content in solid matter of 5.5%, by using the above fluidized bed incineration furnace, the concentrations of nitrogen oxides (NO_x) in the waste gas are as follows: 65 ppm when pulverized coal is added, and 1,230 ppm when pulverized coal is not added.

SEWAGE TREATMENT AND DISPOSAL SYSTEMS

Fluidized Bed Furnace for Aiding Transportation of Partly Dehydrated Sludge

According to U.S. Patent 3,941,065 a fluidized bed furnace is designed such that the transportation costs for the sludge are minimized without causing an excessive throughput in the fluidized bed. This advantage is obtained by providing a furnace roof above the combustion chamber which is substantially horizontal and which is provided with an opening for attaching a post combustion chamber thereabove which has a substantially smaller cross section than the combustion chamber. A supply connection for secondary combustion air is connected into the post combustion chamber. The furnace roof is also provided with another opening for feeding the sludge downwardly into the combustion zone.

With this arrangement of the fluidized bed furnace, the sludge to be burned is transported only to the level of the furnace roof. From this level, the feeding is possible without the use of a gravity tube so that the expense for such a tube and its replacement during the operation are avoided. The distribution of the feed in sludge within the combustion chamber is such that the contact with the fluidized bed in which the supply sludge is dried, pulverized and partly burned and subsequently gasified and incinerated is approximately central and relatively uniform.

The relatively voluminous combustion chamber of the process ensures an extensive burning out of the combustible gases escaping from the fluidized bed and air is supplied at the bottom to maintain the bed. The post combustion takes place in a relatively small post combustion chamber. Since the flow velocity is higher in the post combustion chamber because it has a substantially smaller cross section than the main combustion chamber, the preponderant part of the ashes is entrained there.

With the reduction of sludge transportation costs, it is particularly advantageous to provide sludge treatment devices on the roof of the furnace so that only at this location is the sludge brought into a state more suitable for feeding. Consequently, the sludge can be easily pumped in a still relatively very fluid state to the roof level at which pumping may be easily carried out since the roof level is at a height which is not above the normal pumping height at atmospheric pressure. By appropriate further treatment in centrifuges, filter presses or the like, the sludge can be dehydrated to the necessary extent and thereby made suitable for feeding directly at the furnace roof.

Since the difficulties in the transportation of sludge which is feedable from above are largely eliminated, the post combustion chamber can be designed, as to its length, as a recuperator, so that a large part of the heat produced by the sludge combustion may be recuperated. This is of great importance particularly for sludge combustion because the water content allowable in the sludge to be burned can be higher the more the combustion air supplied into the fluidized bed and the combustion chamber is preheated. Thus, the fluidized bed furnace simultaneously provides conditions for an appreciable reduction of the sludge dehydration costs.

By making the cross section of the combustion chamber substantially larger than the post combustion chamber, and by also designing the chambers so that the combustion chamber has semicylindrical ends and intermediate rectangular portions, an optimal design is effected. Therefore, the usual circular cross section of a fluidized bed may be abandoned.

The central rectangular portion of the combustion chamber may be elongated or made relatively short and the inlet of the post combustion chamber is provided either in the zone of the end half-cylindrical portions or in the center of the rectangle. In the case where the post combustion chamber connects into the center of the central rectangular portion, a feed opening for the sludge may be provided in the zone of each of the two semicylindrical portions so that, even with an elongated fluidized bed, a uniform contact is ensured. Oval or even triangular cross sections may likewise be provided for the combustion chamber so that the conditions for a regular operation of the fluidized bed can be obtained also in larger furnace units.

Partly Dehydrated Sludge Combustion

U.S. Patent 3,939,782 describes a fluidized bed furnace constructed so that as little as possible of the useful bed material is lost during the discharge of coarse particles. In addition, the construction is such that coarse particles will not be discharged prior to being sufficiently treated in the fluidized bed. Also, the discharge of the coarse particles is effected by a construction which is simple in design and great importance is attached to the possibility of reestablishing the operational capacity as soon as possible in case of disturbances. This importance is due to the fact that the storage property and possibility of sludges to be burned is limited while, at the same time, they are produced continually as municipal or industrial sewage sludges. Therefore, as far as possible, the continuous operation must not be interrupted.

The furnace bottom has a plurality of step portions at different levels, which levels vary in a descending order from the inlet toward the discharge opening of the furnace. In a particularly advantageous embodiment of the process, the bottom is subdivided into individual steps which extend at different levels and transversely to the connection line between the discharge opening and the side remote therefrom. Coarser particles which are not as easily fluidized as the fill of refractory grains provided for the fluidized bed are thus accumulated near the discharge opening. This happens also in cases where the coarse particles are relatively light in themselves.

The time necessary for the accumulation of the coarse particles near the discharge opening can be influenced by appropriately dimensioning the level of differences of the indivudual step portions as well as by correspondingly sloping the step surfaces. A particularly advantageous treatment of the coarse sludge particles and a minimized loss by discharge of the useful bed material is obtained by providing a level difference between the adjacent steps of approximately 20 mm. An additional advantage is to provide an inclination of the steps up to 3°.

In order to insure sludge treatment which is as extensive as possible, particularly with sludges containing coarse particles to be discharged, it is useful to provide a furnace feed opening near the portions of the bottom extending at the highest level. In such a construction, the sludge to be treated passes through the fluidized bed which, in itself, is intended for vertical operation substantially in the horizontal direction until the residual coarse particles are accumulated near the discharge opening. For this kind of treatment in the horizontal direction, the fluidized bed is quite suitable.

Quench System for Biochemically Treated Sewage Sludge Fluidized Incinerator

In employing a fluidized bed incinerator for the treatment of the sludge obtained in biotreatment of chemical process wastes, a serious problem is the case of those wastes in which the sludge contains a large amount of fusible salts. The conventional incinerator provides for injection of sludge feed, quench water and air into the top of the incinerator vessel. It was found that the injection of quench water into the top of the vessel led to the agglomeration of the sand and ash entering the vapor removal duct and to frequent rapid plugging of the duct, requiring shutdown of the incinerator and laborious clean-up of the vapor removal system.

U.S. Patent 3,994,244 describes a quench system for a fluidized bed incinerator that prevents frequent plugging of the incinerator vapor effluent removal system. Mechanically, the process consists of a vapor off-take conduit, an optional prequench spray nozzle projecting through the off-take conduit into the incinerator, several optional quench water lines delivering water to a full circle lip on the off-take conduit for distribution, a second conduit, at approximately right angles to the off-take conduit, which communicates with the solids separation and vapor take-off system, main quench water spray nozzles which inject water into the mouth of the second conduit and water supply lines to the nozzles.

The process consists of a method of quenching the vapor effluent and entrained solids from a fluidized bed incinerator by withdrawing the vapors through an optionally water-wetted conduit and reducing their temperature by water quench immediately upon their exit from that conduit, the quench system being designed to avoid injection of quench water into the incinerator vessel itself except for a small amount which may be injected to precool the effluent stream as it is withdrawn from the incinerator vessel.

Figure 12.1 shows a single incinerator vessel **11** with appurtenant equipment. A bed **12** of solid particles is maintained in the lower portion of the vessel. The bed is fluidized by passing gases, preferably air, through it. Air is supplied via line **13** and chambers **14** and **15** and enters the fluidized bed through a grate **16**. During startup of the incinerator, methane is also injected through line **17**; chamber **14** then serves as a combustion chamber to provide hot gases for heating the fluidized bed to operating temperature. When the bed is in operation, a liquid fuel such as kerosene is normally injected through line **18** directly into the fluidized bed.

The arrangement which was employed prior to this process is illustrated by dashed lines in Figure 12.1. Sludge feed was injected into the top of the incinerator vessel through line **21** and quench water through line **22** terminating in multiple nozzles **23**. Air was injected through line **24**, joining the sludge feed and aiding in its atomization through nozzle **25**. In this system, it was found that the heating of the ash above its fusion point caused rapid plugging of the vapor effluent line.

In the process incinerator, sludge feed from line **31** may be injected into the fluidized bed at several levels through lines **32, 33** and **34**, as desired. The lines shown as dashed lines in the drawing are removed. Effluent vapor, containing the combustion gases and entrained solids, is withdrawn through the vapor withdrawal system which consists of conduits **35** and **36**, venturi scrubber **37** in which additional water is injected into the vapors, line **38** and scrubber **39** in which remaining solids are removed from the vapor stream. Substantially solids-free vapor is exhausted through line **40** and a slurry of solids in water leaves the scrubber through line **41** to settler **42**, from which a relatively clear stream of water is removed through line **43** and a slurry is pumped back through line **44** to vessel **37**, together with makeup water from line **45**.

This process is concerned primarily with the quenching of the vapor effluent stream from the incinerator. As shown schematically, quench water is injected into the vapor stream leaving conduit **35** and entering conduit **36**, through a quench water line **46** equipped with at least one nozzle **47**.

FIGURE 12.1: QUENCH SYSTEM FOR FLUIDIZED INCINERATOR

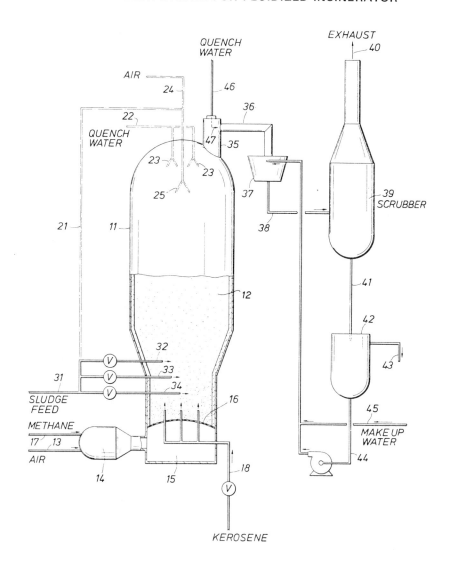

Source: U.S. Patent 3,994,244

Production of Active Carbon

U.S. Patent 3,887,461 describes a process of removing solids from sewage, subjecting these solids to pyrolysis, thereby producing active carbon and char, and using both of these materials as adsorbent agents in the treatment of the sewage or other wastewater.

After the sewage has been dewatered, the solids which are filtered out of the sewage in the form of cake are conveyed to a dryer where they are at least partially dried by hot gases. A screw feeder introduces the sludge cake into a fluidized bed reactor where it is flash heated and carbonized and where it simultaneously disintegrates into small particles, most of which will pass through 16 mesh (U.S. Standard Sieve) and some of which will pass through 325 mesh.

Hot flue gas is supplied to the fluidized bed reactor by means of a blower and the velocity of the gas flowing through the reactor is sufficient to maintain the carbonizing sludge particles in a fluidized state. Any suitable fuel can be combusted in the burner. The temperature within the reactor is held within the range of 500° to 1000°C, and the gas contains little free oxygen, so that the sludge is subjected to pyrolysis.

The only oxygen contained in the combustion gases is the excess air supplied to the burner in order to maintain optimum combustion within this burner (optimum combustion being accomplished with a slight excess of air). Thus the amount of free oxygen contained in the combustion gases flowing through the reactor would be within the range of 0.1 to 2% by volume. The sludge particles, after having their residual moisture flashed off and most of their pyrolytic volatile products driven off, are in the form of active carbon and char, which does not require further conventional activation (selective oxidation) over extended periods, typically one to several hours, at elevated temperature, to exhibit useful adsorptive capacity.

The gases discharged from the fluidized bed pass through a wet scrubber (or other dry collector) where the fine carbon particles carried along in the gas stream are separated out. The separated carbon particles are relatively pure active carbon, typically having (BET) surface area of 300 m^2/g or more. The carbon slurry from the scrubber is recycled to the sewage system and introduced therein to adsorb impurities from the sewage and to markedly improve the quality of primary effluent. The activated carbon can be introduced downstream of the tank, or can alternatively be introduced upstream of, or directly into the tank, and still be effective.

Inert Fluidized Bed

A process described in U.S. Patent 3,319,587 is concerned with direct combustion of organic waste material having a high moisture content, under specially controlled conditions. The waste matter is fed continuously into a body of hot inert fluidized granular material having heat-storing and heat-radiating capabilities, e.g., sand.

The fluidized or teeter state of this granular material is maintained in the combustion chamber by the combustion air being forced upward therethrough.

Thus, the air pressure should be sufficient to keep the sand particles in teeter which in turn depends upon the inventory of inert particles to be thus maintained in depth. The devices for introducing the wet waste material into the combustion chamber should provide an adequate seal against that pressure.

Under the foregoing conditions, and with certain controls of the combustion process itself, direct and complete combustion of the wet waste is attainable with the result that the stack gases discharging from the combustion chamber comprise only inert and innocuous combustion gases along with water vapor, carrying with them the residual ash and inert fines.

To this end, the controls of this combustion process should maintain a high enough combustion temperature in the body of fluidized material, while maintaining a low excess of combustion air. However, optimum efficiency of such a combustion operation further depends upon achieving the combustion of the organic waste within the confines of the body of hot inert fluidized material, which presents the problem of minimizing or eliminating the afterburning of waste material and/or of its distillation gases that might occur in the freeboard space above the body of the fluidized granular material.

According to the process, under the above conditions afterburning is minimized by extruding the high moisture waste material tightly compacted into the lower zone of the body of hot inert fluidized material maintained in the combustion chamber.

Thus, the waste material is delivered from an extrusion device in the form of plugs or chunks compacted to a density approximately equal to the density of or greater than the surrounding body of fluidized material. Gradually these chunks become more and more fragmented, finally disintegrated, and destroyed by combustion, with evaporation of the water taking place concurrently within the confines of the body of inert fluidized material.

In related work U.S. Patent 3,319,586 describes a process which involves effecting the combustion of the moist raw sludge in a bed of inert hot granular material maintained in a state of fluidization which is the formation of a dense and turbulent suspension of particles in an upflowing stream of treatment gas termed a "turbulent layer" or fluidized bed.

Combustion thus conducted under the conditions of the process will reduce the organic solids in the sludge to stable gaseous products, with a large portion of the heat liberated by the combustion of the sludge fuel used to flash off the moisture in the sludge while also heating the combustion gases to the bed temperatures ensuring complete odorless combustion.

There is hereby provided a method where a concentrated sludge, for example, in the range of 25 to 35% solids concentration, is fed directly into a combustion chamber where, in the presence of the suspended inert particles of hot inert auxiliary material, a complete and odorless combustion is maintainable and readily controllable so that the combustion gases will be kept odor-free. To that end, this process provides a combustion chamber wherein a quantity of inert granular material having high heat storage, as well as high heat-radiating capability, for example, graded sand, is maintained in a heated fluidized state.

Screen Deliquefier

A process described in U.S. Patent 3,375,794 relates to an internal screen de-liquefier for pressure conduits carrying moist fibrous material, and is particularly useful with screw or progressive cavity feeding of dewatered waste organic material into an incandescent fluidized inert material bed reactor as described above in U.S. Patent 3,319,587.

Pressure-Cooked Sludge as Feed

A process described in U.S. Patent 3,580,193 relates to disposing of sewage or other waste sludges by combustion by first heat treating the sludge at an elevated temperature and pressure for an extended period of time. This treatment results in a sludge which may be mechanically dewatered to a cake having a much higher solids content and which is much more friable and easily handled than conventionally thickened and mechanically dewatered sludge.

By combining the pretreatment of sludge with means by which the sludge is fed continuously to dewatering means such as a solid bowl centrifuge or a rotary filter and with continuous feed to a fluid bed combustion chamber, an extremely compact and efficient thermal disposal system results. This sytem requires minimum space, no preheating of combustion air, and requires no auxiliary fuel during normal operation.

The product of the sludge pressure-cooked after mechanical dewatering is unlike other dewatered sludges which are plastic and cohesive. This pressure-cooked sludge after mechanical dewatering is of a dry friable nature, substantially free from odor, which readily crumbles into small pieces which are especially suited to extremely rapid and substantially complete combustion in a fluidized bed of inert granular material. Such material may, despite its high solids content, be fed by a screw feed, and due to its crumbly nature may be also broken up into small pieces and blown into the combustion chamber.

The most striking characteristic of the cake resulting from heat treating most sewage sludges in accordance with this process is the friable crumbly nature of the cake which renders it easily broken into small pieces, somewhat granular in structure. This greatly aids feeding and permits a choice of feeding methods such as milling and air conveying or blowing to the combustion chamber or screw feeding, and promotes distribution and even burning in the combustion chamber. It is particularly advantageous when fed to a fluidized bed combustion system because of its rapid and substantially uniform dispersal in the fluidized bed.

FURNACE DESIGNS AND MODIFICATIONS

Stepped Distributor Plate

To aid in solids feed distribution, U.S. Patent 3,915,657 describes a fluidized bed reactor characterized by a distributor plate, in which there are at least two fluidizing surfaces. The solids feed to the reactor is preferably discharged below the upper fluidizing surface; any of the well-known methods of heating the reactor can be employed. Thus, the fluidizing gas can be preheated, the reactor plenum chamber or reactor fluidizing chamber can be heated directly by internal firing or indirectly through the use of heated jackets and the like, i.e., any convenient combination of heating methods can be employed.

The fluidized bed reactor is particularly useful for disposition of combustible waste or sewage sludge. In a typical application, the combustible waste is pumped or screw conveyed into the reactor at a point below the first fluidizing surface. The fluid bed heat transfer material, typically silica sand, is preheated to about 1500°F and is maintained at that temperature during operation of the reactor principally as a result of the heat of combustion of the combustibles in the waste feed.

In cases where there is a high water content in the waste feed, supplementary fuel is often added to the reactor in the vicinity of the waste feed inlet so as to maintain the desired combustion temperature. The fluidized bed reactor can be readily operated substantially in the absence of air to pyrolyze combustible waste or sewage sludge.

In operation, there is an advantage in performance as each fluidizing surface is supplied with fluidizing gas from a separate plenum chamber. The use of separate plenum chambers for each fluidizing surface makes the performance of the surface independent of the pressure drop at each fluidizing surface, and permits independent control of the intensity of fluidization for each fluidizing surface.

In a typical application for incineration of sludge from a sewage treatment plant, the lower fluidizing surfaces of the distributor plate and the portion of the fluidized bed defined within the volume encompassed between the base fluidizing surface and the uppermost fluidizing surface can be operated at a lower temperature than the portion of the bed and reactor above the upper fluidizing surface. This approach reduces the possibility of charring or plugging of the feed to the reactor at the feed inlets to the reactor due to lower temperatures at the feed inlets.

In a sewage incineration application, the temperature within the volume defined by the distributor plate is preferably operated at a temperature which is 50° to 100°F less than the average temperature in the fluidizing chamber. The temperature is principally controlled within the volume defined by the distributor plate by controlling the rate of fluidizing gas to the plate. At low flow rates, the lower portion of the fluidized bed, i.e., within the volume of the distributor plate, is most viscous in behavior and this reduces the amount of heat brought into this section of the fluidized bed by conduction from the hotter upper section of the bed.

Feed and Exhaust Modifications for Disposal System

U.S. Patent 3,922,975 describes a process for consuming solid waste using a fluid bed reactor wherein low-fuel-content, heavy particles are first removed from the solid waste and the high-fuel-content, lighter fraction particles introduced into the fluid bed reactor with minimum air turbulence and as densely as possible.

In accordance with the process, heavy particles are separated from the lighter particle fraction of solid waste in an air classification stage and the lighter particle fraction carried via a conveying gas to an inertial separation stage wherein the lighter particle fraction is separated from the conveying gas stream which in large part is returned to the air classification stage. The separated lighter particle fraction solid waste material is conveyed from the storage area to an air-lock feed valve into a fluid conduit for delivery into the fluid bed reactor.

To overcome the problems existing in such systems for efficient operation using high air flow and still avoiding loss of the particulate matter from the fluid bed and excessive contamination of the exhaust gases with fly ash and other air pollutants, U.S. Patent 3,921,544 describes a process for particle collection in the exhaust of the system.

The process provides a method and apparatus for separating the granular fluid bed material and fly ash entrained with exhaust gases from a fluid bed reactor for efficient operation and permissible operation under existing air pollution standards. The separated granular material is returned to the fluid bed.

Fly ash which is separated from the exhaust gas stream with the granular material in the first stage of inertial separation is then separated from the granular material and carried to residue storage.

The first and second stage inertial separation is conducted in heat exchange with the combustion air that is being directed to fluidize the granular bed in the combustion chamber both to cool the inertial separation stages and to preheat the combustion air.

Coarse Particle Discharger

U.S. Patent 3,910,208 provides a fluidized bed furnace having a discharge device designed so that coarse particles can be removed from the fluidized bed which would be too large and hard for treatment by pulverization in the bed into flue dust and which would be such that they disturb the operation when they are left in the material of the fluidized bed. The process makes it possible to burn sludges in a fluidized bed which normally would be unsuitable for such treatment.

The discharge opening from the furnace is located directly adjacent to the bottom which supports the fluidized bed, and it is connected to a rotary drum having a diameter comparable to the discharge opening diameter, and which is large enough to permit the passage of the coarsest particles which are expected. The inside of the drum wall is designed as a screw having a pitch which is provided with blade elements which are spaced apart sufficiently to accommodate the size of the coarsest occurring particles.

There is a very large discharge opening in the wall of the fluidized bed furnace and the diameter, for example, is approximately 400 mm. Through such an opening, even the coarsest residues could be removed.

In order to retain the bed material to the necessary extent, a rotary drum is provided which turns at a very low speed. The threads or blades of the screw part of the drum prevent the bed material from flowing out, but they permit the secure pickup of the coarse particles which are collected at the free end of the screw, for example, in a receiving chamber or shaft.

Since the portion of the screw within the blades or the central passage of the screw is substantially smaller than the coarse particles, only a very small amount of the bed material can flow out freely during the operation and the screw threads discharge only a quantity of bed material corresponding to the rotational speed of the screw. Thus, in this discharge device, the drive of the drum is advantageously started only after a large amount of coarse particles has accumulated at the inside of the furnace.

In the remaining time, the outflow of the bed material through the free inner passage of the screw can be prevented by a shut-off gate provided between the discharge opening and the drum. It is even possible, for example, to stop the drum as soon as the last of a sequence of coarse particles has left the free discharge end and to reverse the drive of the drum so that the bed material which has unintentionally passed between the screw threads is returned into the furnace. Because of this construction, the central passage of the drum remains free and a correspondingly higher speed is used for the reverse drive of the drum.

Air and Fuel Supply Lance

U.S. Patent 3,910,209 describes a lance in which oil may be fed to a fluidized bed of burning sludge material and, if necessary, may provide an additional heating device which may be operated for a long period of time without disturbance and which also permits an undisturbed operation with waste oil. Disturbances may occur in such a lance primarily by causing a cracking of the fuel at high temperatures.

In intermittently operated lances, disturbances of this kind are very frequent. It has been found that such disturbances can be avoided by maintaining a very high flow velocity within the lance. This favorable effect can be further supported by measures which keep the lance cool.

In accordance with the process, the lance is provided with a further connection which leads through an adjustable valve to a second compressed air line so that additional compressed air may be fed into the lance as soon as the feeding of the oil drops off. It is particularly advantageous if the lance is disposed so as to extend approximately horizontally through the furnace wall in the zone of the fluidized bed.

It is possible to add compressed air to the oil feed which must be effected under a pressure corresponding to the pressure of the compressed air, in order to maintain a very high velocity during the operation and, particularly, also during periods when the heating of the oil is reduced or even stopped.

In the cases where the oil feeding is stopped, the lance is operated only with air. With the intended use of the lance in a fluidized bed, the flow velocity may be chosen very high without running the risk of breaking of the flame because a sufficient volume of combustion air is present in the fluidized bed for the fuel oil fed in. Consequently, there is no need for a special design of the furnace end of the lance and, in fact, this end may be formed with a straight cut-off end portion.

Instead of the fuel oil, other fuels, such as coal dust or gaseous hydrocarbons, may also be used. It is particularly advantageous for this purpose to use waste oil which must be destroyed anyway. Due to the possibility of securely avoiding a cracking of the added fuels in the hot zone of the lance, the lance ensures an undisturbed operation in any case and over long periods of time.

Should solid particles pass into the lance tube along with the material to be burned, or should cracking occur due to an error of operation or supply disturbance, a push rod may be used which may be pushed through a plug connection after the plug is removed so that it moves the entire length of the lance and into the furnace. This means that the lance may be easily cleaned of any extraneous matter.

Figure 12.2 is a partial sectional view of a fluidized bed furnace with a lance constructed in accordance with the process. Referring to the figure, the process comprises a fluidized bed furnace, generally designated **50**, which includes a bottom **1** over which a fluidized bed is formed up to a level, for example, at **52** and which includes a combustion chamber **54** defined above fluidized bed **52** and below a furnace roof **56**. Air is supplied through a wind box **58** and nozzles **60** through the bottom **1** in order to maintain the fluidized bed **62**. The furnace wall means includes a furnace side wall **2** having a refractory lining and an outer sheet metal shell **3**.

In accordance with the process, a tubular socket connection **4** extends into the side wall **2** above the bottom **1** in the area of the fluidized bed **62** and it is supported on a support **7** of the furnace. The tubular socket includes an end flange **4a** which is secured to an end flange **5** of an outer or first tubular member **9** which extends at its inner end into the fluidized bed **62** and is sealed with the connection socket **4**. The outer tube is provided with a connection **10** to a first compressed air supply **64** and, in addition, its extreme outer end **10a** is provided with an opening through which an inner tubular member or lance **8** is directed.

Lance **8** is supported on spacers **11** at a spaced location from the interior of the outer tube or jacket **9**. The outer tube includes an opening **12** for a pressure-measuring device which is connected to a pressure pick-off **16**. The rear or outer end of the lance tube **8** is provided with a second compressed air connection **13** which extends through a control valve **20** having a control motor **19** and to a compressed air supply **21**. A fuel connection **14** is also located to the rear of the second compressed air connection **13** at the tube **8**.

The pressure which is sensed by the pressure pick-up **16** is fed to a regulator or controller **17** for the control motor **19** of valve **20** and it supplies compressed air through the second compressed air connection **13** in accordance with the pressure which is sensed. The controller is also provided with an input **18** for controlling the pressure in the lance in accordance with a preset condition so that if the pressure sensed varies from this, the valve **20** will be either opened or closed accordingly. The rear end of the lance tube **8** is closed by a blind plug **15** which may be removed in order to permit the insertion of a push rod for cleaning the entire tube.

FIGURE 12.2: AIR AND FUEL SUPPLY LANCE

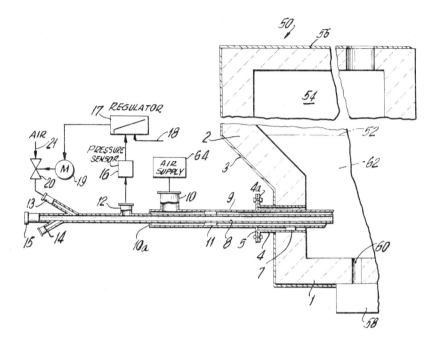

Source: U.S. Patent 3,910,209

The process makes it possible to maintain a permanently sufficient flow velocity and prevent the cracking of the fuel oil and, in addition, it makes it possible to control the supply of the fuel oil due to the aspirating or injector action of the compressed air which is fed in through the conduit connection **13** ahead of the fuel connection **14**. This control for the fuel oil is more effective than a nozzle control which would be subjected to variations due to viscosity and other property changes in the fuel being fed.

The pressure value which is obtained by the pick-off **16** is converted to an electrical signal which is applied to the regulator **17**. The desired value input **18** may be varied in accordance with operating characteristics and this value normally corresponds to the required minimal velocity at which cracking would be avoided. As soon as the pressure drops below this value, the regulator **17** starts the motor **19** to open valve **20** for the connection to the second compressed air supply **21**.

Multiple Fluidized Beds for Iron Ore Reduction

U.S. Patent 3,985,547 describes a process for direct reduction of ores in fluid beds inexpensively and with no possibility of the metal produced being contaminated with sulfur.

According to the process, ores and especially iron ores, are reduced directly in a multistage fluid bed reactor, in which the reducing and fluidizing gases are the products of partial combustion of methane with oxygen, the gases being introduced into an intermediate zone of the reducing tower above the stage or stages where final metallization occurs.

The process takes place at pressures of between 1 and 5 atmospheres; methane and oxygen in substoichiometric proportions are introduced into a combustion chamber, the outlet of which is connected to the reactor in an intermediate zone, for instance, between the last and the next-to-the-last reduction beds. In the reduction chamber the methane reacts with the oxygen to produce carbon monoxide, hydrogen and water and a small percentage of carbon dioxide. Part of the methane, generally less than 10%, remains unburned and circulates in the reactors without causing trouble.

The gas thus produced, together with the gas coming from the lower stages of metallization, passes through the upper beds where the ore is in a low-reduced state. In this way the amounts of CO_2 and H_2O present do not hinder the progress of this phase of the reduction process. Once the reducing gas has passed through the reactor and emerged from its top, the dust and the water it has picked up are separated out and the CO_2 removed, after which the gas is reheated and sent to the last reducing bed.

In this way, the reducing gas, composed essentially of CO and H_2, is fed precisely in the zone where the presence of a pure reducing gas is most necessary, thus completing the ore reduction process. The carbon black produced in the partial combustion chamber can be removed before it enters through the grating of the overlying fluidized bed.

Referring to Figure 12.3, methane and oxygen are sent via pipes **12** and **13** to combustion chamber **6**, where the methane is partly burned with the oxygen, the oxygen being fed in substoichiometric quantities. The hot gases from the combustion are stripped of carbon black and sent to chamber **7** disposed between beds **2** and **3** of reactor **1**. The percentage composition, by volume, of the gases is H_2: 45–53, CO: 28–33, CO_2: 8–4, H_2O: 9–4, CH_4: 5–2. They are mixed in chamber **7** with the gases recycled through chamber **14**, which have passed through the final metallization bed **2** completing reduction of the burden.

The mixture thus obtained passes upwards to fluidize and reduce the ore contained in beds **3**, **4** and **5** respectively, becoming richer in CO_2 and H_2O and poorer in CO and H_2 at each stage of reduction. The gas mixture leaves the reactor through pipe **15**, is dedusted in cyclone **16** and then passes to the heat exchanger **8** where it is cooled.

From here the gas is sent to units **9** and **10** where water and carbon dioxide are stripped, respectively. Then some of the cleaned gas is bled off at **18**, while the remainder is recycled through unit **8** where it is preheated by indirect heat exchange with the gas leaving reactor **1** through pipe **15**. From **8**, the gas is sent to the final heating furnace **11**, from which it reenters reactor **1** at **14**.

The ore is introduced to reactor **1** through feeder **17** and flows down through beds **5**, **4**, **3** and **2** in countercurrent with the fluidizing and reducing gas. From the last reducing bed **2**, which may also be considered the final metallization bed, the product is discharged at a degree of reduction that exceeds 93%.

FIGURE 12.3: MULTIPLE FLUIDIZED BEDS FOR IRON ORE REDUCTION

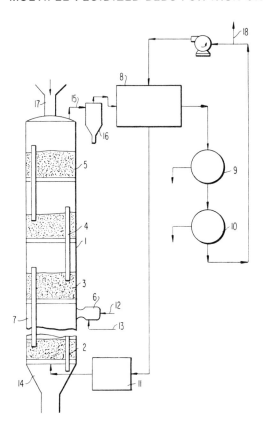

Source: U.S. Patent 3,985,547

Fluidized Bed for Nongaseous Combustible Material

U.S. Patent 3,970,011 describes apparatus suitable for burning in a fluidized bed a nongaseous combustible material to produce a solid incombustible residue. The apparatus comprises a combustion chamber and a support adapted to support a bed of particles in the combustion chamber and to admit gas to the bed so that the bed can be fluidized. Also included are a first discharge means for discharging nongaseous material into the bed, a second discharge means for discharging solids from the bed and feed means for mixing a gaseous fuel and air and feeding the mixture into the bed through the support at a rate such as to fluidize at least a part of the bed.

One problem which arises when nongaseous combustible material is to be burned in a fluidized bed is that of raising the temperature of the bed to the minimum temperature at which combustion of the material can occur.

With this process, the temperature of the bed can be raised by feeding a mixture of gaseous fuel and air into the bed and igniting the mixture in the combustion chamber. Initially, the gaseous fuel will burn above the bed and impart heat to the bed. As the temperature of the bed rises, combustion will spread into the bed. The feeding of a gaseous mixture of fuel and air into the bed enables complete combustion of the fuel to be ensured of suitable proportions of gaseous fuel and air, without any necessity for feeding a large excess of air into the bed with the fuel.

The apparatus may also have a passageway through which the gaseous mixture flows to the support, the support forming one boundary of the passageway and a wall of thermally conductive material forming an opposite boundary so that heat can be extracted from the passageway through the wall. The wall may be provided with fins which project away from the passageway and means may be provided for directing a fluid coolant over the fins.

This arrangement reduces the risk of combustion of the fuel occurring prematurely, i.e., before the gaseous mixture of fuel and air has passed into the bed through the support.

The apparatus may be so arranged that, when the bed is fluidized by gas fed into the bed through the support, a general flow of particles is established along a circulatory path.

Such a general flow of particles assists distribution of the nongaseous combustible material throughout the bed and reduces the risk of a part of the bed becoming substantially cooler than the remainder of the bed. Thus, the combustible material discharged into the bed by the first discharge means may be wet, partly combustible material such as refuse.

Secondary Bed Elutriation Arrestor of Labyrinth Construction

U.S. Patents 3,882,798 and 3,818,846 describe a fluid bed disposal method and apparatus including a secondary bed/elutriation arrestor obstruction of labyrinth construction in the fluid bed reactor having lower and upper vertically spaced-apart perforate retention plates between which are positioned at least first and second vertically spaced-apart grates with the grate bars of the grates staggered with respect to one another. The lower surface of the bars of the lower perforate plate is described as stepped between at least two horizontal surface locations. At least two fluid bed/elutriation arrestor labyrinth obstructions spaced apart above the bed of granular material are used.

While the process is well suited for the pollution-free disposal of solid waste with the possible additional simultaneous disposal of various liquid materials, one configuration of the process is useful in a municipal solid waste disposal plant where the solid waste can be used as a fuel to dispose of high water content municipal sewage sludge from a given population segment.

One of the more difficult materials to burn in a fluid bed arrestor is aluminum-laden solid waste with the aluminum in the form of aluminum foil, flip-top can lids and the like. During combustion of this type of material, molten nodules of aluminum are formed and upon impact with one another will oxidize, resulting in slagging, clinkering and ash buildup.

The addition of the secondary bed/elutriation arrestor provides a means of breaking apart these nodules as they flow through the labyrinth obstruction, depending upon the configuration of the obstruction. It has been found during tests that a labyrinth obstruction in the form of a plurality of large spheres such as of ceramic positioned between a pair of spaced-apart perforate plates results in aluminum oxide buildup and consequent unacceptable use when burning aluminum-laden solid waste.

In this process, the fluidizing air flows through the bed material particles under carefully controlled conditions, chief among these conditions being the requirement that the air velocity through the bed, and hence the pressure drop, be greater than the value required to support the bed weight and less than the value required to sweep the particles out of the bed. In addition, the bed consists of particles within a suitable range of size, shape and density. When these conditions are all satisfied, the stationary bed of particles will have expanded and the bed particles will exist in a fluidized state.

If the movement of one specific particle could be observed, it would be seen to undergo a continuous, turbulent motion and would wander throughout the bed in a random manner. Viewed as a whole, the dynamic condition of the fluidized bed resembles a tank of boiling water in the sense that there is considerable turbulence and bursting of bubbles at the surface. This dynamic characteristic gives the fluidized bed its unique advantages.

Cascade Addition of Sludge

U.S. Patent 3,736,886 illustrates that it is possible to vastly increase the efficiency of a fluidized bed furnace for the combustion of sludge, especially a furnace operating with an expanded fluidized bed, in which a distillation of sludge is permitted to cascade or free fall onto the bed and an afterburning of the volatile components of the sludge is performed above the inlet of the latter but within the fluidized bed chamber.

The clarifier sludge is supplied to a fluidized bed furnace in the form of a liquid, paste or solid (generally a semisolid or a flowable solid), in which a fluidized bed is maintained in a burning state with the aid of oxygen-containing fluidizing gases. At least a portion of the sludge (or other waste to be burned) is introduced into the chamber at a point above the highest combustion zone of the expanded fluidized bed (i.e., above the flame at the top of the fluidized bed), the point at which the sludge is fed to the furnace being disposed below or at a sufficient spacing from the exhaust gas outlet, that within the fluidized bed chamber, there is effected an afterburning of volatile constituents of the sludge.

Surprisingly, the simple free fall or cascade of the sludge, or a portion thereof, to the expanded fluidized bed from above may result in a distillation of volatile constituents regardless of the original consistency of the sludge, these volatile constituents interacting with the gases rising from the fluidized bed to provide the afterburning mentioned earlier. The fact that afterburning is carried out completely within the fluidized bed chamber and between the inlet for the sludge and the gas outlet, provides maximum utilization of the heat value of the materials.

The fluidized bed furnace illustrated in Figure 12.4 comprises a refractory-lined, steel-encased shaft 1 provided with a grate 2 defining the bottom of a fluidized bed chamber 1a. Above the grate there is maintained an expanded fluidized bed 3 of inert fluidized solids, e.g., quartz sand, at which combustion of volatile components of a sludge and burnable solids is carried out.

Below the grate, there is formed a windowbox 2a into which an oxygen-containing fluidizing gas stream is introduced from a combustion chamber 5. The latter may be fed with atmospheric air from a compressor or blower 4 through an inlet 5a when only air is to be introduced or when air is to be supplied to the fluidized bed in combination with combustion gases. A valve 4a controls the proportion of air supplied for this purpose. Another valve 4b controls the air supplied to a burner 6 opening into the combustion chamber and supplied with fuel at a line 6a. The fuel introduced here may be gas or oil.

It is apparent that heat can be supplied to the fluidized bed by sustaining combustion in chamber 5 and introducing a mixture of hot combustion gases and air to the fluidized bed. At any rate the fluidizing gas must be rich in oxygen or must contain oxygen in a stoichiometric excess over any combustible materials introduced by the fluidizing gas. Another portion of the fluidizing air, controlled by a valve 4c, may be fed at 7a through a heat exchanger 7 and returned at 7b to the windowbox. The heat exchanger is located above the gas outlet 13 at the upper end of the combustion chamber 1a.

It has been found to be advantageous, from time to time, to introduce unheated air or oxygen-enriched air to the combustion chamber above the fluidized bed or into the fluidized bed by lances represented at 8a, 8b and 8c, the latter lances serving to supply preheated air to the system. The gases distributed to these lances via supply lines 8 through valves are represented at 4d, 4e and 4f. Fuel may be introduced through the lances 8c and 8d as well. Spaced above the top 3a of the fluidized bed, which is maintained approximately in line with the primary sludge inlet 9, by a distance 11, is the secondary sludge inlet 10.

The inlet is located 6 meters above the grate and about 4.5 meters above the fluidized bed which may have a high ranging between 1 and 1.5 meters. The pipe 10, moreover, should be located somewhat above (say 0.5 to 1 meter) the top of the combustion zone. Above this pipe, at least at its mouth 10a, there is located the exhaust gas outlet 13 at a spacing of at least 3 meters and represented by the dimension 12. Within this afterburning zone, the gas velocity is maintained at or below 3 m/sec. A temperature control 13a at the outlet regulates the operating parameters of the burner and gas feed to maintain the exhaust gas temperature in the range of 800° to 900°C.

FIGURE 12.4: CASCADE ADDITION OF SLUDGE

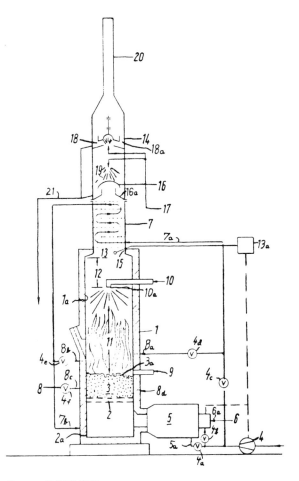

Source: U.S. Patent 3,736,886

The hot exhaust gases discharged through the outlet **13** pass at **15** into the heat exchanger **7** in which sensible heat is transferred to gases which are used to feed the bed. Thereafter, the exhaust gases traverse a scrubber **14**, e.g., a radial flow scrubber as shown in Figure 12.4, and then enter a chimney or stack **20** which is advantageously mounted directly upon the furnace. Water is sprayed into the scrubber as shown at **17**, **18** and **19** and a shield is provided at **16** of the bell type. The shield cooperates with a cone **16a** and an annular gap **18a** is provided at the scrubber **18** for additional separation of impurities. The shield **16** and the gap **18a** are designed so that the scrubbing water is properly drained at **21** with the resulting seals acting as an explosion vent. The annular gap is adjusted in response to pressure differential to maintain a superatmospheric pressure within the combustion chamber and insure a constant scrubbing effect even when the combustion rate and the exhaust gas rate vary. The scrubbing of the exhaust gas effectively removes sulfur trioxide and hydrogen chloride where alkaline scrubbing water is employed.

As a specific example of the process, with the outlet 10a located 3 meters below the gas outlet 15 but 6 meters above the grate 2, with a fluidized bed of quartz sand maintained at a height of about 1.5 meters and a gap of 1 meter between the top of the combustion zone and the secondary waste inlet 10a, the temperature maintained at the gas outlet 15 is about 800° to 900°C. The system is used to burn refinery sludge consisting of 95% by weight water and 5% oil and industrial solids.

Without a secondary combustion as illustrated, 300 to 500 kg of refinery sludge could be processed per hour per square meter of the grate area. Using a secondary flow of 30 to 60% of the total flow, it was possible to increase the combustion rate to 600 to 1,000 kg of sludge per hour per square meter of the grate area. The fluidizing air is heated to a temperature of 450° to 600°C and the temperature can be controlled to maintain a level of 800° to 900°C at outlet 13 by feeding fuel into the system at the lances 8c and 8d.

Multistage Dryer Above Furnace

U.S. Patent 3,772,998 describes a sludge-burning method and apparatus in which a fluidized bed forms the combustion zone and is surmounted by a multistage drying and preheating zone such that the exhaust gases of the combustion zone rise directly through the multistage zone above the fluidized bed.

The multistage drying and preheating zone is in the form of a multiple-hearth furnace with annular supporting surfaces, each associated with one or more rotary brakes or agitators which continuously break up the layer on the respective stage and keep the sludge moving downwardly from stage to stage until a dried, preheated, comminuted solid material is distributed over the entire cross section of the fluidized bed. Hence it is essential that the fluidizing gases contain oxygen to effect a combustion in the bed, that the wastes are comminuted and transformed into a particulate or other subdivided form during drying and preheating, and that the comminuted matter is distributed over the grate area of the furnace.

In the multiple-hearth furnace the annular, vertically spaced, axially superposed hearths are cantilevered from the inner wall of the housing of the drying zone and define a central space in which a shaft is rotatable, this shaft carrying a plurality of radial arms at each stage, the arms having depending plate-like rake teeth for scraping the deposit on the annular platform overhung thereby. The teeth may be provided at various inclinations to the rotation circles defined by the arms, i.e., lie along tangents or chords to induce a movement of the comminuted waste spirally inwardly or outwardly depending upon the location of the outlet for each platform.

The platforms are alternately provided with outlets along the inner periphery or the outer periphery so that, for example, the comminuted drying material may be induced to flow inwardly at one annular platform, is caused to cascade downwardly onto an intercepting inner peripheral portion of the next lower platform, is induced to move outwardly along this latter platform by the associated rake or comminution arms, and is caused to cascade downwardly along the outer periphery of this platform onto still another platform underlying it.

Since the fluidized bed furnace forms the combustion zone and is surrounded by the multiple-hearth furnace, a transfer conduit is provided for recycling the waste gases of the multiple-hearth furnace into the fluidized bed furnace. The action of the rake teeth produces a comminuted material which is highly combustible and easily strewn across the grate of the fluidized bed, or, more specifically, across the full cross section of the fluidized bed furnace with uniform distribution.

Nozzle Feed System

U.S. Patent 3,709,170 describes a sludge feed system for a reactor having a nozzle, and a conduit communicating with the outside of the reactor and the nozzle for feeding sludge to the nozzle under pressure, the nozzle having a mouth in communication with the interior of the reactor for forming the sludge into a thin layer which breaks up in the reactor.

Referring to Figures 12.5a and 12.5b, enclosed reactor 10 has a lower cylindrical portion 11, a central conical section 12 and an upper cylindrical portion 13. The described reactor portions comprise an exterior metal shell with a suitable refractory material lining the interior surface thereof, particularly if the reactor is to be employed in high-temperature reactions or treatments. The bottom 14 of the reactor and the top 15 are of similar construction to the reactor vertical walled portions. The reactor is typical of the type used in fluidized bed treatments and particularly in the combustion of waste materials.

A constriction plate 16 spans the horizontal interior area defined by the lower cylindrical portion of the reactor support. The space between the constriction plate and the reactor bottom constitutes air box 18. A plurality of vertical tubes 17, in the constriction plate, communicate at their lower ends with the air box. The upper ends of the tubes, which extend above the constriction plate, terminate in horizontally located small tubular sections 19 through which air is expelled from the air box under pressure to fluidize particulate bed 20 located in the reactor. Conduit 21 feeds air or hot gases or whatever gas is appropriate to the air box. Conduit 22 in the upper part of the reactor is used to withdraw gases and products of combustion from the reactor.

Nozzle 30 is positioned so that its mouth is in communication with the interior of the reactor. As shown in Figures 12.5a and 12.5b, the nozzle mouth is so positioned in the lower portion of the reactor as to be slightly above the top surface of the fluidized bed 20. However, the nozzle can be so located as to deliver sludge being fed from the mouth directly into the fluidized bed.

It is advisable, although not essential in all cases, to provide means to regulate the temperature around and in the nozzle. This can be accomplished by suitable jacket means provided adjacent to the nozzle through which a suitable fluid can be circulated for temperature control of the nozzle.

FIGURE 12.5: NOZZLE FEED SYSTEM

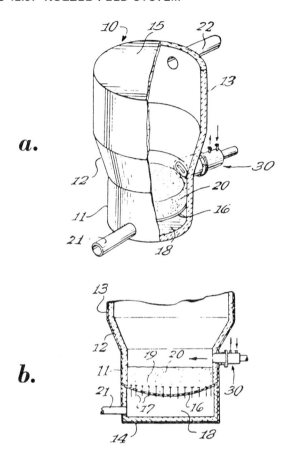

(a) Isometric View, Partially Broken Away, Showing a Reactor
Provided with a Nozzle for Feeding Sludge in a Thin Layer
to the Interior

(b) Vertical Sectional View Through the Lower Part of the
Reactor of Figure 12.5a

Source: U.S. Patent 3,709,170

To facilitate insertion of the nozzle in the reactor and to aid its removal for
servicing and repair, an opening is provided in the reactor wall to receive the
nozzle and suitable attaching means. Thus, a plate is positioned normal to and
joined to the outer wall of the cooling jacket. Insulation material is positioned
around the inner end of the nozzle and is suitably secured thereto and to the
plate.

It is considered advisable to dimension the nozzle mouth to have a width at least five times, and better at least ten times, greater than the mouth height to shape the sludge into a suitably thin layer. More specifically, it is usually appropriate to eject the sludge as a layer of no more than 2-inch thickness, and advisably of no more than 1-inch thickness. For optimum results it is advisable to have the layer no more than ½ inch thick and about 12 to 15 inches wide.

A specific nozzle for feeding 1,700 lb/hr of wet sewage sludge of about 28 to 33% solids in water to a fluidized bed reactor can have a transition element circular inlet of 3-inch diameter and a rectangular mouth ½ inch high and 14 inches wide. The sludge layer ejected from such a nozzle is not self-supporting and breaks up as it leaves the nozzle mouth. The scrubbing action of the fluidized bed will break the sludge pieces into smaller particles which will completely burn.

Distributor Plate Arrangement

U.S. Patent 3,772,999 describes a fluidized bed apparatus for combusting or reacting mixtures of combustible and noncombustible waste matter with a distributor plate having sides sloping toward the bottom of the apparatus and having an inlet for solid feed in the side of the distributor plate. A pipe extends downwardly from the base of the distributor plate and functions to capture noncombustibles in the solid feed. A lock is provided which permits noncombustibles to be removed from the apparatus without interruption of operation.

U.S. Patent 3,776,150 describes a fluidized bed system for pyrolysis or incineration of solid wastes. The solids feed is force-fed to a fluidized bed apparatus having a conically shaped distributor plate and a first internal chamber above the distributor plate and a smaller second internal chamber connected to the first chamber and positioned directly below the first chamber.

Solid waste is fed into the apparatus by a feeder which seals the feeder from the apparatus at a point spaced from hot zone of the apparatus. The system has fluidizing gas inlets which are designed so that the fluidized bed in the first internal chamber is more highly fluidized than the fluidized bed in the second internal chamber. A lock in the second internal chamber permits removal of nonreacted materials without interruption in operation of the system.

Dryer and Heat Exchange System

U.S. Patent 3,779,181 describes a heat exchange system which permits the economic disposal of a slurry of combustible waste materials by drying the waste materials prior to reaction (i.e., incineration or pyrolysis), utilizing off-gases from the reactor in which the combustible materials are reacted to effect the drying.

The fluid bed dryer comprises a drying chamber and a heating jacket contiguous with the drying chamber. The drying chamber and the heating jacket are each adapted to contain a bed of fluidizable particulate heat transfer material positioned in their lower portions and both the drying chamber and the heating jacket have a fluid gas inlet and an exhaust gas outlet. The drying chamber also has a feed inlet and a feed outlet. Preferably, the feed outlet is vertically spaced from the feed inlet and a conveyor means extends from the feed outlet into the drying chamber for transferring dried solids from the drying chamber to the feed outlet.

The heat exchange system comprises in combination:

    (a) a fluid bed dryer such as described above,

    (b) means for conveying a slurry comprising liquid and combustible waste solids into the drying chamber of the dryer,

    (c) a reactor or incinerator having a feed inlet, a reaction section and a vapor exhaust port,

    (d) conveyor means for transferring solids from the fluid bed dryer feed outlet to the feed inlet of the reactor,

    (e) conduit means connecting the vapor exhaust port of the reactor to the gas inlet of the heating jacket of the fluid bed dryer, and

    (f) means for providing fluidizing gas to both the reactor and the drying chamber of the fluid bed dryer.

The heat exchange system may include a conduit communicatively connecting the exhaust gas outlet of the drying chamber of the fluid bed dryer with the reactor. The reactor or incinerator is preferably a fluidized bed reactor.

In this system, the hot off-gases resulting from incineration of combustible waste solids are used to fluidize heat transfer material in the heating jacket of the dryer. The cross-sectional area of the heating jacket must be sized so that the incinerator off-gas is sufficient to form a well-agitated fluidized solids bed. Typically 0.5 to 1.5 ft/sec superficial velocity of gas in the jacket is suitable. A high heat transfer coefficient results between the wall of the fluidized bed heating jacket and the wall of the fluidized bed drying chamber.

The system can result in a heat transfer coefficient of about 50 Btu/hr-ft^2-$^\circ$F on each side of the drying chamber wall (assuming the fluidized bed in the dryer is well agitated). The heat transfer system also eliminates heat exchanger fouling because of the lower temperature at the heated walls of drying chamber, as a result of increased heat transfer, and reduced contact of the solids being dried with the heat exchange wall. Another advantage to the process is that odorous volatiles can be deodorized by afterburning in the vapor-disengaging section of the incinerator and made substantially odor-free before being discharged into the atmosphere.

The operating temperature of the fluidized bed dryer for drying waste feeds containing high water concentrations, i.e., up to about 90% water, is 212°F. Air is provided to the fluid bed dryer from a blower in an amount to maintain a state of minimum fluidization (about 0.5 ft/sec superficial vapor velocity) in the bed of the dryer without the presence of waste feed. This fluidization level insures that the waste feed material can be pumped into the drying chamber without difficulty. Vaporization of water and volatile constituents in the waste feed provides additional fluidizing gas and results in a well-agitated fluidized solids bed (about 2 ft/sec superficial vapor velocity is desirable).

In actual tests performed on drying of a paunch manure feed containing about 82% by weight water, steam, air and other gases are discharged from the dryer exhaust and piped to the vapor disengagement section of an incinerator, where these gases are combined with the incinerator combustion gases which are at elevated temperature on the order of about 1500°F. The combined off-gas equilibrates to a temperature of about 1350°F. The off-gas enters the base of the heating jacket at about 1350°F and is discharged from the heating jacket at about 700°F. The water content of the slurry is reduced by about 60%. The cooled off-gas is piped to a water scrubbing system for removal of particulates and soluble contaminants and discharged to the atmosphere.

High-Melting Inert Particles

In a process described in U.S. Patent 3,411,465 moist waste material is incinerated in fluidized beds of higher-melting inert particles in an approximately cylindrical vessel where a slow stirring is applied to the fluidized bed, and if desired, dehydration of feedstock is effected in that vessel by passing combusted gas upward through a dehydrating bed charged with moist feed material while applying slow stirring. The gas is flowed upward at a sufficiently high rate relative to that of the particles to prevent downflow of water, thereby causing free water to become present at the upper end of the vessel, and withdrawing free water.

In Figure 12.6a, an example is shown where the apparatus of the process is applied to the incineration of fowl excretions. The apparatus comprises an upright cylinder. Air for combustion is supplied, at a temperature in the vicinity of normal temperature or after being preheated, from the bottom portion of the apparatus through an inlet **101** and through a perforated plate **102** to the fluidized bed **103** in which solid particles such as sand grains are fluidized. **104** represents the peripheral wall of the incineration furnace. **105** represents an ignition burner or fluid inlet which is used to heat the interior of the apparatus to the combustion temperature.

The fowl excretions which are to be incinerated are fed through a supply inlet **106** into the apparatus. While the excretions descend through the fluidized bed, they are contacted by the high-temperature combusted gas and by the sand grains and thus they are heated and dried, eventually combusted to ashes.

The fluidized bed is held at a high combustion temperature by virtue of the combustion temperature which is produced during the foregoing process. In the event that there is want of calories, a combustion gas may be directly introduced into the bed through a gas inlet **105** by means of a burner or any other appropriate means. **107** represents a combusted gas exhaust pipe. **108** represents an afterburner for depriving the combusted gas of its odor. **109** represents a supply inlet of sand grains to the fluidized bed.

It is preferred that the fowl excretions which constitute the material to be incinerated are continuously fed into the apparatus in their dispersed state. Since it may occur, however, that the excretions are fed into the apparatus in a mass of considerable size dropping directly on the perforated plate **102** where they may accumulate to block the openings of the plate, there is provided obstructing means **110** in the form of a coarse grate while gentle stirring is performed in the area immediately above this obstructing means by stirring vanes **111'** to thereby disperse the masses and to avoid the occurrence of the aforesaid blocking of the openings of the perforated plate.

FIGURE 12.6: FLUIDIZED BED INCINERATOR

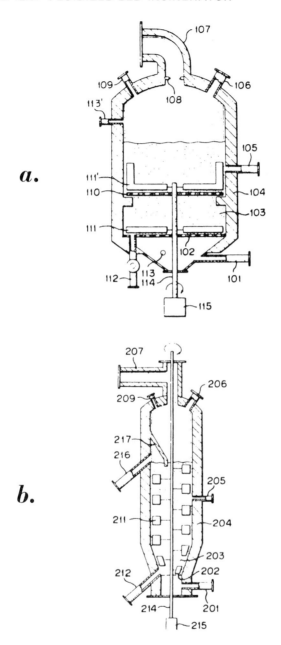

(a) Incineration of Fowl Excretions
(b) Incineration of Sludge from Pulp Factory

Source: U.S. Patent 3,411,465

The masses of excretions which have passed through the spaces in the grate **110** downward and which have not dispersed into smaller clods are dried, destroyed and dispersed in the highly heated fluidized bed and are eventually combusted to ashes. The area immediately above the perforated plate **102** is also gently stirred to avoid adherence of the particles to the plate by the stirring vanes **111** which are similar to those described above, and these vanes are mounted to the periphery of the common rotary shaft **114**.

Depending on the type of the material to be incinerated, the combination of the grate and the stirring vanes **111'** or the meshes of the grate may be appropriately modified. Furthermore, the combination of the grate and the stirring vanes may be provided in plural sets, one set being located above the other. In some instances, one or both of them may be omitted as required. The grate, the perforated plate, and the stirring vanes are constantly polished by being contacted by the fluidized particles. Therefore, no blocking of the meshes of the grate after the particles of waste materials have dropped downward therethrough will occur.

Such materials as fowl excretions which are to be incinerated often contain small shells and small pebbles which would heap up on the perforated plate. These obstacles are removed together with a part of sand grains through a discharge outlet **112**. This removal of pebbles or the like may be performed jointly through a withdrawing outlet provided at an appropriate portion of the apparatus such as the region where the grate is located. The sand grains which have been taken out from the apparatus are screened to remove the foreign elements, and the segregated sand grains may be mixed with fresh sand grains and supplied to the fluidized bed **103** through the supply inlet **109** of sand grains.

Numeral **115** represents a low-speed motor which is used to rotate the stirrer shaft **114** at low speed. The vanes **111** and **111'** which are fixed to the periphery of this shaft are preferably so angled as to work to take up particles and send them upward since such shape of the vanes will serve to effect smooth operation of the apparatus continuously for an extended period and to thereby save the power.

113 and **113'** represent holes for measuring the pressure. Besides these, a plurality of thermocouples for the measuring of temperature are inserted in the wall of the apparatus at appropriate sites thereof. Those fine powders and ashes which scatter outside the system from the outlet **107** are collected by a powder separator such as cyclone. For the sake of recovery of heat, a boiler or other appropriate means may be provided.

Figure 12.6b illustrates an example of the apparatus which is applied to the treatment of condensed sludge from the liquid-form waste discharged from a pulp factory. The reference numerals appearing on this figure whose last two digits are the same as the last two digits of the reference numerals appearing in Figure 12.6a identify corresponding parts. Such sludge contains small pieces of wood, barks, wood meal and fibrous materials or organic substances such as dissolved lignin or inorganic materials.

Inorganic materials are represented by sodium sulfite and sodium carbonate. This example shows how such inorganic materials are collected as solid particles.

This is effected by feeding condensed sludge into an upright cylindrical apparatus in which solid particles of inorganic particles of appropriate size are fluidized and which has a conically shaped bottom structure, whereby perfect combustion of all that is combustible takes place.

In case the material to be treated is in the form of suspension in a liquid, it may be supplied in atomized state into the apparatus through a supply inlet located at the top, 206, or it may be injected directly into the fluidized bed at the inlet 205 through the peripheral wall 204. Coarse particles are drained from the draining outlet 212 located at the bottom of the apparatus, while relatively fine particles are drained from a draining outlet 216 provided near the surface of the fluidized bed.

In addition to these outlets, there may be provided a particle-draining outlet at an arbitrary level of the fluidized bed. A part of the sludge introduced into the apparatus is immediately dried and combusted and reduced to fine particles of inorganic substances alone, which will either stay in the bed by being involved in the fluidized bed or will be discharged together with the combusted gas outside the system from the outlet 207 and will be collected by cyclone so as to be returned to the bed again.

These fine particles of inorganic substances will act as cores and collect the sludge onto their surfaces where the deposited sludge is dried and combusted. Thus, the particles of inorganic substances will gradually grow into larger particles and they are discharged locally outside the system from the particle-draining outlets.

By arranging so that a small air current flows near the outlets against the direction in which the particles are discharged, this air current will serve to cool and at the same time to classify the particles, so that particles of inorganic substances having relatively uniform size will be collected easily. 217 is a partition wall to prevent by-passing of particles around the bed.

At a temperature lower than the melting point of particles, sintering and fusing of particles are completely avoided by the stirring of the fluidized bed. Therefore, by employing as high a temperature as possible and by thus holding the bed in the state of reducing atmosphere it is possible to collect the inorganic substances in the form of reclaimed chemical particles for being used again. In some cases, however, consideration and care may have to be given to hold the structural materials of the apparatus below the permissible temperature by passing water or the like to cool the stirrer shaft and the inside portions of the vanes.

Fluidized beds are usually of an average density of 1.0 g/cc or more, so that even when wood pieces or the like are introduced into the apparatus they are dried and combusted while they are floating on the surface of the bed. Thus, wood pieces or like materials will never sink deep into the bed. If small pebbles or nails, which have a larger specific gravity, are mingled into the apparatus, they will be taken up by the current produced by the vanes rotating at low speed immediately above the perforated plate 202 located in the bottom portion, and thus they are transferred to the outlet 212 wherefrom they are discharged together with coarse particles.

U.S. Patent 3,515,381 describes a method for burning sludge by introducing the sludge in dispersed form from the top of the combustion chamber into a fluidized layer of inert particles traversed by a constant quantity of oxidizing fluidizing agent.

The temperature of the fluidized layer of inert particles is maintained at constant value to insure the combustion of the product to be treated and the temperature of the zone above the fluidized layer is maintained at a constant value less than the sintering temperature of the ash formed by regulating the speed of introduction of the dispersed sludge. A temperature-controlling fluid is injected into the system over the whole section of the chamber immediately under the fluidized layer.

Multiple Bed Fluidized Solids Burner

U.S. Patent 3,306,236 describes a burner for waste materials, i.e., for wastes which may contain gases, solids and/or liquids. The burner operates on the fluidized-solids principle. Fine particles of refractory material are used as heat carriers. A multiple bed arrangement is employed with one dense fluid solid bed preferably arranged above another. Both of these zones are fluidized by passing gases, preferably air, upward through them. Elastic fluids, which may be either gases or vapors or both, are used to keep the dense beds fluidized. They are introduced to the lower zone and preferably directed into the lower bed. As the vapors and liquids are consumed or vaporized, etc., residual solids such as ash, or carbon in some cases, will be deposited upon the more refractory solids, which comprise the major part of the original solids in the lower vessel.

The vessel may be constructed as a single unit with walls of suitable material, such as steel, lined with refractory brick or cement, etc. Alternatively, separate vessels may be superimposed. The temperature T_1 in the lower bed is sufficiently high, after the system gets into operation, to insure gasification in some way, e.g., at least vaporization or partial cracking, etc., of all the waste products except their ash or perhaps their carbon content.

For example, a good working temperature is about 1000°F. It may be less, but preferably will not be below 800°F nor more than 1200°F in most cases. At these temperatures in the lower fluidized bed, any waste material supplied thereto, whether in liquid or solid form, will either be vaporized, or cracked or burned, at least to some extent.

Generally speaking, ash will remain as the chief solid component and it is deposited on or mixed with the refractory particles constituting the original bed. Heat for the lower fluidized bed is supplied primarily by a stream of solids flowing down a downcomer from the upper level. Gases passing upward from the interface of the lower bed ordinarily will not be completely consumed or oxidized at this point. They will contain highly undesirable atmospheric pollutants of at least a nuisance type. They may even be dangerous in some situations, if they were released in such condition.

They may contain odorous compounds and/or incomplete combustion products such as carbon monoxide or hydrogen, H_2S, etc. Additional fuel is injected to burn these materials completely. The hot gases including the complete combustion products then pass through the upper bed.

They impart some but not all of their heat. Therefore, the upper bed will have a temperature T_3 substantially higher than that of the lower bed, though less than T_2 between them. Finally, the gases emerging from the upper bed have a temperature T_4 at which they are passed through the solids separating device such as a cyclone. Separated solids return to the upper bed through a down-spout in the conventional manner.

The arrangement, in essence, is such that the gases rising from a lower zone through the upper fluidized bed give up a certain amount of heat, which heat equals, of course, the sensible heat of the stream of solids circulating back to the lower bed. The solids which rise, leaving the lower bed at a temperature T_1, for example, of 1000°F will reach the higher temperature T_3 in the upper vessel. The difference is essentially the heat applied to vaporization, combustion, crack-ing, etc., of the waste products fed into the lower bed.

The primary point of control of this sytem is the temperature of the lower bed. It is controlled eventually by firing supplementary fuel above the bed in suitable quantity, the heat being imparted first and mainly to the solids in the bed above, which reach the temperature T_3. In many cases the supplementary fuel added need be only very small in quantity. When there is enough fuel value in the waste products being consumed to substantially meet all heat requirements, fuel will not be required at all. Generally speaking, however, with waste products such as sewage which contain a great deal of water, often around 90%, fuel will be required. The fuel feed rate is generally controlled in accordance with the water content of the waste products being consumed.

The rate of air flow is generally kept substantially constant and sufficient to insure good fluidization and proper mixing in the two beds as well as solids cir-culation between them. The rate of refractory solids cycling between the two beds also is preferably maintained substantially constant.

Surplus solid materials, if and as they accumulate, may be withdrawn. It may be necessary, under some conditions, to supplement or occasionally replenish the solids which constitute the fluidized beds. For this purpose refractory solids of suitable type such as granules of metal oxides, metallic particles, ceramic par-ticles, etc., and mixtures thereof may be added.

Since the temperature in the lower bed is generally rather moderate, the grid or grate at the bottom usually can be made of relatively inexpensive materials. Preferably an alloy is used which is reasonably resistant to oxidation at the op-erating temperatures. In cases where the waste solids include pieces of metal or other solids too large to fluidize, these may accumulate on the grate and may require periodic removal therefrom.

This may be done either through a suitable cleanout or by shutting down the system and removing the grid, etc. Generally speaking, the upper bed will not require a grate in the sense of a support for massive solids. Normally, the first dense fluid solids bed is arranged beneath the second and is at lower temperature, the temperature between the two beds being higher than either. In some cases, the beds need not be superimposed in this manner. The second bed supplies heat to the first by solids circulation.

BIBLIOGRAPHY

The following reports used in the preparation of this book are available from:

National Technical Information Service
U.S. Department of Commerce
5285 Port Royal Road
Springfield, Virginia 22151

ANL/ES-CEN-1008   G.J. Vogel, W.M. Swift, J.F. Lenc, J. Montagna, S. Saxena, P.T. Cunningham, W. Hubble, and A.A. Jonke. *A Development Program on Pressurized Fluidized-Bed Combustion.* Quarterly Report July 1, 1974-Oct. 1, 1974. Oct. 1974.

BNL-19308   S. Chalchal, T.V. Sheehan, and M. Steinberg. *Coal Combustion and Desulfurization in a Rotating Fluidized Bed Reactor.* October 1974.

CONF-751213-3   J.S. Wilson and D.W. Gillmore. *Preliminary Report on Fluid-Bed Combustion of Anthracite Wastes,* 1975.

CONF-760402-4   A.P. Fraas. *Application of the Fluidized Bed Coal Combustion System to the Production of Electric Power and Process Heat.* AIChE, Kansas City, Missouri, April 11-14, 1976.

ERDA-76-69   A. Himmelblau and J. Norton. *Concept for Fluidized Bed Combustion of Consol Char Using a Closed-Cycle Helium Power Plant with an Estimate of the Price of Electric Power.* Final Report April 1976.

FE-1514-42   S. Lemezis. *Advanced Coal Gasification System for Electric Power Generation, Multiple-Fluidized-Bed Coal Gasification System: Conceptual Design and Cost Estimate.* June 1975.

ICP-1088   R.E. Schindler. *Emergency Fluidization and Bed Removal During Fluidized-Bed Solidification of Commercial Wastes.* April 1976. ERDA Contract E(10-1)-1375 S-72-1.

PB 206 892   E.G. Davis, I.L. Feld and J.H. Brown. *Combustion Disposal of Manure Wastes and Utilization of the Residue.* January 1972.

PB 211 323   *Sewage Sludge Incineration.* Environmental Protection Agency Report No. EPA-R2-72-040. August 1972.

PB 212 960   D.H. Archer, D.L. Keairns, J.R. Hamm, et al. *Evaluation of the Fluidized Bed Combustion Process. Volume II. Technical Evaluation.* November 1971.

PB 231 162   D.L. Keairns, D.H. Archer, J.R. Hamm, R.A. Newby, E.P. O'Neill, J.R. Smith and W.C. Yang. *Evaluation of the Fluidized-Bed Combustion Process. Volume I. Pressurized-Bed Combustion Process Development and Evaluation.* December 1973.

PB 231 977   *Proceedings of Third International Conference on Fluidized-Bed Combustion.* December 1973.

PB 233 101   D.L. Keairns, D.H. Archer, R.A. Newby, E.P. O'Neill and E.J. Vidt. *Evaluation of the Fluidized-Bed Combustion Process. Volume IV. Fluidized-Bed Oil Gasification/Desulfurization.* December 1973.

PB 234 343   A.H. Bagnulo, J.W. Bishop, S. Ehrlich, and E.B. Robison. *Development of Coal Fired Fluidized Bed Boilers. Volume I.* July 1971.

PB 234 344   J.W. Bishop, S. Ehrlich and J.S. Gordon. *Development of Coal Fired Fluidized-Bed Boilers. Volume II.* Sept. 1972.

PB 237 028   *Energy Conversion from Coal Utilizing CPU-400 Technology.* November 1974.

PB 237 366   G.J. Vogel, M. Haas, W. Swift, J. Riha, C.B. Schoffstoll, J. Hepperly, and A.A. Jonke. *Reduction of Atmospheric Pollution by the Application of Fluidized-Bed Combustion.* Annual Report July 1972-June 1973. June 1974.

PB 237 754   G.J. Vogel, W.M. Swift, J.F. Lenc, P.T. Cunningham, W.I. Wilson, A.F. Panck, F.G. Teats, and A.A. Jonke. *Reduction of Atmospheric Pollution by the Application of Fluidized-Bed Combustion and Regeneration of Sulfur-Containing Additives.* September 1974.

PB 241 834   D.L. Keairns, R.A. Newby, E.J. Vidt, E.P. O'Neill, C.H. Peterson, C.C. Sun, C.D. Buscaglia, and D.H. Archer.

Fluidized Bed Combustion Process Evaluation. Phase I. Residual Oil Gasification/Desulfurization Demonstration at Atmospheric Pressure. Volume I. Summary. March 1975.

PB 241 835   D.L. Keairns, R.A. Newby, E.J. Vidt, E.P. O'Neill, C.H. Peterson, C.C. Sun, C.D. Buscaglia, and D.H. Archer. *Fluidized Bed Combustion Process Evaluation. Phase I. Residual Oil Gasification/Desulfurization Demonstration at Atmospheric Pressure. Vol II. Appendices.* March 1975.

PB 246 116   D.L. Keairns, D.H. Archer, J.R. Hamm, S.A. Jansson, B.W. Lancaster, E.P. O'Neill, C.H. Peterson, C.C. Sun, E.F. Sverdrup, E.I. Vidt, and W.C. Yang. *Fluidized Bed Combustion Process Evaluation. Phase II. Pressurized Bed Coal Combustion Development.* September 1975.

PB 260 478   R.C. Hoke, R.R. Bertrand, M.S. Nutkis, D.D. Kinzler, L.A. Ruth and M.W. Gregory. *Studies of the Pressurized Fluidized-Bed Coal Combustion Process.* Sept. 1976.

RFP-2016   D.L. Ziegler, A.J. Johnson, and L.J. Meik. *Fluid Bed Incineration.* June 8, 1973.

RFP-2471   D.L. Ziegler. *Fluidized Bed Incineration of Radioactive Waste.* May 14, 1976.

The following U.S. Patents used in the preparation of this book are available from:

Commissioner of Patents and Trademarks
Washington, D.C. 20231

3,306,236   D.L. Campbell; Feb. 28, 1967; assigned to Esso Research and Engineering Company

3,319,586   O.E. Albertson and W.E. Budd; May 16, 1967; assigned to Dorr-Oliver Inc.

3,319,587   O.E. Albertson and W.M.H. Kilmer; May 16, 1967; assigned to Dorr-Oliver Inc.

3,375,794   O.E. Albertson and W.M.H. Kilmer; April 2, 1968; assigned to Dorr-Oliver Inc.

3,411,465   T. Shirai; Nov. 19, 1968

3,515,381   P. Foch; June 2, 1970; assigned to Charbonnages de France, France

3,580,193   R.P. Logan and O.E. Albertson; May 25, 1971; assigned to Dorr-Oliver Inc.

3,709,170	L.R. Van Gelder; January 9, 1973; assigned to Chicago Bridge & Iron Co.
3,736,886	R. Menigat; June 5, 1973; assigned to Metallgesellschaft AG, Germany
3,772,998	R. Menigat; November 20, 1973; assigned to Metallgesellschaft AG, Germany
3,772,999	C.S. Miller, Jr., H.K. Staffin, and R. Staffin; Nov. 20, 1973; assigned to AWT Systems, Inc.
3,776,150	P.R. Evans and D.H. Graham; Dec. 4, 1973; assigned to AWT Systems, Inc.
3,779,181	H.K. Staffin and R. Staffin; Dec. 18, 1973; assigned to AWT Systems, Inc.
3,818,846	R.G. Reese; June 25, 1974; assigned to Combustion Power Co., Inc.
3,882,798	R.G. Reese; May 13, 1975; assigned to Combustion Power Co., Inc.
3,887,461	R.D. Nickerson and H.C. Messman; June 3, 1975; assigned to Combustion Engineering, Inc.
3,888,194	K. Kishigami, H. Kobayashi, T. Sente and K. Sugiyama; June 10, 1975; assigned to Babcock Hitachi KK, Japan
3,907,674	E.J. Roberts and P.A. Angevine; Sept. 23, 1975; assigned to Dorr-Oliver Inc.
3,910,208	E.A. Albrecht, H.W. Oepke and H. Wulfmeier; Oct. 7, 1975; assigned to Rheinstahl AG, Germany
3,910,209	E.A. Albrecht, H.W. Oepke and H. Wulfmeier; Oct. 7, 1975; assigned to Rheinstahl AG, Germany
3,915,657	R. Staffin and R.E. Tkac; Oct. 28, 1975; assigned to Hercules Incorporated
3,916,805	C.D. Kalfadelis and A. Skopp; Nov. 4, 1975; assigned to Exxon Research & Engineering Co.
3,921,543	R. Menigat and W. Fennemann; Nov. 25, 1975; assigned to Metallgesellschaft AG, Germany
3,921,544	R.G. Reese; Nov. 25, 1975; assigned to Combustion Power Co., Inc.
3,922,975	R.G. Reese; Dec. 2, 1975; assigned to Combustion Power Co. Inc.

3,926,129   C.J. Wall; Dec. 16, 1975; assigned to Dorr-Oliver Inc.

3,939,782   E. Albrecht; Feb. 24, 1976; assigned to Rheinstahl AG,
                   Germany

3,941,065   E. Albrecht; March 2, 1976; assigned to Rheinstahl AG,
                   Germany

3,970,011   M.J. Virr and D.E. Elliott; July 20, 1976; assigned to
                   Fluidfire Development Ltd., England

3,985,547   I. Iacotti, G. Malgarini and E. Pasero; Oct. 12, 1976;
                   assigned to Centro Sperimentale Metallurgico, SpA, Italy

3,994,244   W.R. Pledger and J.E. Gwyn; Nov. 30, 1976; assigned to
                   Shell Oil Co.

The following journal articles were also used in the preparation of this book.

P.F. Fennelly, H. Klemm, and R.R. Hall, "Coal
Burns Cleaner in a Fluid Bed," *Environmental
Science & Technology*, Vol 11, No. 3, March 1977,
p. 244-48.

T.J. Hanson and D.J. Swanton, "Fluidisation—
Combustion Related Application," *Technology
Ireland*, Vol 8, No. 9, Dec. 1976, p. 31-34.

K.P. Becker and C.J. Wall, "Fluid Bed Incineration of Wastes," *Chemical Engineering Progress*,
October 1976, p. 61-68.

NOTICE

Nothing contained in this Review shall be construed to constitute a permission or recommendation to practice any invention covered by any patent without a license from the patent owners. Further, neither the author nor the publisher assumes any liability with respect to the use of, or for damages resulting from the use of, any information, apparatus, method or process described in this Review.

Complete copies of the patents described in this book may be obtained at a cost of $0.50 per patent prepaid. Address order to the Commissioner of Patents, U.S. Patent Office, Washington, D.C. 20231.

UNDERGROUND COAL GASIFICATION 1977

by George H. Lamb

Energy Technology Review No. 14

This book presents an overall view of underground coal gasification starting with its history, thereby explaining the theoretical basis for the different methods, the typical processes, including U.S. program studies, the problems encountered, special instrumentation and equipment required, foreign advances, economical aspects and future outlook.

Although *in situ* gasification of coal will not eliminate all energy problems, it appears to have great promise in alleviating some of the problems associated with the production, transportation, and burning of coal.

While not expected to replace coal mining, underground gasification offers an alternative method of bringing the energy value of the coal to the surface. If economically feasible, this clean form of energy could supplement the natural gas supply, particularly in the industrial sector where it could be used also to produce electricity.

Most of the information presented in this book is based on federally funded studies. Here are condensed vital data that are scattered and difficult to pull together. Experimental equipment and set-ups are reviewed and detailed by actual case histories. Each chapter is followed by its own list of references to the most recent literature. A partial and condensed list of contents follows here.

ISBN 0-8155-0670-8

255 pages

ENERGY FROM BIOCONVERSION OF WASTE MATERIALS 1977

by Dorothy J. De Renzo

Energy Technology Review No. 11
Pollution Technology Review No. 33

One of the chief gaseous products of the anaerobic decomposition of organic matter is methane, CH_4. This is how natural gas was formed in prehistoric times along with other fossil fuels.

By applying this principle today in environmentally acceptable fashion it is possible to bioconvert municipal solid sewage, animal manure, agricultural and other organic wastes into substitute natural gas (95% CH_4). In its simplest essentials the process consists of loading the material into a digester (a closed tank with a gas outlet). Given favorable thermal and chemical conditions, the appropriate biological processes will then take their course.

The bioconversion of waste materials to methane provides at least partial solutions not only to the energy problem, but also to the solid waste disposal problem. The harvesting of heretofore undesirable vegetations, such as algae, water hyacinths, and kelp as "energy crops" offers unconventional opportunities for supplementary utilization of natural resources.

This book describes practical methods for the bioconversion of waste matter. It is based on reports of academic and industrial research teams working under government contracts. A partial and condensed table of contents follows here. Chapter headings and important subtitles are given.

Note: Each chapter is followed by bibliographic reference lists in order to provide the reader with easy access to further information on these timely topics.

ISBN 0-8155-0656-2

223 pages

HOW TO SAVE ENERGY AND CUT COSTS IN EXISTING INDUSTRIAL AND COMMERCIAL BUILDINGS 1976
An Energy Conservation Manual

by Fred S. Dubin, Harold L. Mindell and Selwyn Bloome

Energy Technology Review No. 10

This manual offers guidelines for an organized approach toward conserving energy through more efficient utilization and the concomitant reduction of losses and waste.

The current tight supply of fuels and energy is unprecedented in the U.S.A. and other countries, and this situation is expected to continue for many years. Never before has there been as pressing a need for the efficient use of fuels and energy in all forms.

Most of the energy savings will result from planned systematic identification of, and action on, conservation opportunities.

Part I of this manual is directed primarily to owners, occupants, and operators of buildings. It identifies a wide range of opportunities and options to save energy and operating costs through proper operation and maintenance. It also includes minor modifications to the building and mechanical and electrical systems which can be carried out promptly with little, if any, investment costs.

Part II is intended for engineers, architects, and skilled building operators who are responsible for analyzing, devising, and implementing comprehensive energy conservation programs. Such programs involve additional and more complex measures than those in **Part I**. The investment is usually recovered through demonstrably lower operating expenses and much greater energy savings.

A partial and much condensed table of contents follows here:

Much of the technology required to achieve energy savings is already available. Current research is providing refinements and evaluating new techniques that can help to curb the waste inherent in yesteryear's designs. The principal need is to get the available technology, described here, into widespread use.

ISBN 0-8155-0638-4

725 pages

DEEP COAL MINING
Waste Disposal Technology 1976

by William S. Doyle

Pollution Technology Review No. 28

The environmental impact of coal mining has been recognized and described since medieval times. Drainage from active and abandoned mines has always been a source of river pollution, although former generations could afford to ignore it.

Coal refuse banks also are a source of acid mine drainage (AMD) and of silt. In addition these unsightly refuse banks are susceptible to spontaneous combustion and can smolder for months or years, thus contributing substantially to air pollution.

This book, based on government reports and important recent U.S. patents, discusses methods to prevent and control pollution associated with deep mining of coal. Various processes for neutralization of AMD, the role of iron oxidation and methods for oxidizing the iron are described. Treatments of AMD with soil or by ion exchange and reverse osmosis are detailed. One chapter describes the treatment of AMD combined with municipal wastewater.

Also discussed is the underground disposal of mine wastes, the necessary sealing compounds, as well as the extinguishing of burning refuse banks and their reclamation with the aid of planned vegetation.

A partial and condensed table of contents follows here.

Chapter headings are given and some of the more important subtitles are included.

ISBN 0-8155-0619-8

392 pages

COAL CONVERSION TECHNOLOGY 1976

by I. Howard-Smith and G. J. Werner

Chemical Process Technology Review No. 66

The industrialized countries of the world are showing renewed and increasing activity in the area of synthetic fuels production from coal. Existing commercial coal conversion processes, such as the Fischer-Tropsch synthesis, the destructive distillation of coal, methanation etc. resulting in liquid and gaseous fuels, are again becoming economically competitive with petroleum.

The many processes and techniques of coal conversion have as a basic concept the transmutation of coal into forms acceptable to our transportation and heating equipments. To accomplish this, high-sulfur coals must be desulfurized, high-ash coals must be demineralized, and, most important of all, solid coal must be depolymerized into liquid and gaseous products that can be ignited and burned with facility.

Over 100 processes, in various stages of development, are available to carry out these procedures and are discussed fully in this book.

Originally prepared as an "in-house" report for the Millmerran Coal Pty. Ltd. of Brisbane, Australia, it is intended to provide readily accessible and concise data on all major activities in coal conversion technology. A special feature are the flow charts which clearly illustrate all the principal processes. A partial and very condensed table of contents follows here. Chapter headings are given, and some of the more important subtitles are also included.

ISBN 0-8155-0614-7

STRIP MINING OF COAL
Environmental Solutions 1976

by William S. Doyle

Pollution Technology Review No. 27

Late in 1974 the Office of Research & Development of the U.S. Environmental Protection Agency made the statistically well founded projection that for the remainder of the 20th century surface-mined coal will have to account for over 50% of our nation's production of this fuel.

Strip mining can be done responsibly without permanent damage to land and water. Technology exists for effective reclamation of mined lands, and such reclamation is being performed in some areas. Authorities emphasize the importance of planned premining (fully discussed in this book) and point out that reclamation is less costly and more effective when integrated with the mining operation.

This book, based on 19 government reports issued from 1967 through 1974, describes surface mining of coal, land use and methods, land reclamation technology plus sediment and erosion control. Acid mine drainage, its sources, prevention and correction, as well as the mechanism of reclaiming acid strip mine lakes are discussed. Specific studies on revegetation, use of spoil amendments etc. are included. A whole chapter is devoted to West Germany's approach to the problem of strip-mined lands. The final two chapters are devoted to costs, economics, and financing.

A partial and condensed table of contents follows here. Chapter headings are given, followed by examples of important subtitles.

ISBN 0-8155-0611-2

HYDROGEN TECHNOLOGY
FOR ENERGY 1976

by David A. Mathis

Energy Technology Review No. 9

Hydrogen is attractive as a fuel because it is abundant, relatively inexpensive and ecologically clean. When hydrogen is burned in air, it forms water vapor only, there are no solid combustion residues and no soot particles or noxious gases to contaminate the atmosphere.

The use of hydrogen as a universal fuel necessitates development of methods for storing, handling, and transferring, and these are the main subjects treated in this book. Future volumes in this series will be reserved for the technology of the myriad of hydrogen production schemes now under consideration viz. electrolysis of seawater by ocean-derived thermal energy, by solar cells, certain algae, by nuclear powered direct thermochemical conversion, or from municipal waste, etc.

The first chapter describes the hydrogen economy and suggests how it can be integrated into the USA energy system. The next three chapters are concerned with handling the various forms of hydrogen: gas, liquid, and solid (in the form of metal hydrides). The fifth chapter describes some of the work which has been done or is under way in using hydrogen as a fuel or in an energy storage system. Another chapter delves into safety and the political, socioeconomic, and environmental implications of a hydrogen economy. The final chapter provides a list of hydrogen technology experts and includes a brief description of each individual's expertise in the various aspects of hydrogen technology.

A partial and condensed table of contents follows here.

ISBN 0-8155-0629-5

285 pages